Ralf G. Longwitz

Micro Discharges, Gas Ionization and Detection

AF061204

Ralf G. Longwitz

Micro Discharges, Gas Ionization and Detection

A Study of Gas Ionization in a Glow Discharge and Development of a Micro Gas Ionizer for Gas Detection and Analysis

Südwestdeutscher Verlag für Hochschulschriften

Impressum/Imprint (nur für Deutschland/ only for Germany)
Bibliografische Information der Deutschen Nationalbibliothek: Die Deutsche Nationalbibliothek verzeichnet diese Publikation in der Deutschen Nationalbibliografie; detaillierte bibliografische Daten sind im Internet über http://dnb.d-nb.de abrufbar.
Alle in diesem Buch genannten Marken und Produktnamen unterliegen warenzeichen-, marken- oder patentrechtlichem Schutz bzw. sind Warenzeichen oder eingetragene Warenzeichen der jeweiligen Inhaber. Die Wiedergabe von Marken, Produktnamen, Gebrauchsnamen, Handelsnamen, Warenbezeichnungen u.s.w. in diesem Werk berechtigt auch ohne besondere Kennzeichnung nicht zu der Annahme, dass solche Namen im Sinne der Warenzeichen- und Markenschutzgesetzgebung als frei zu betrachten wären und daher von jedermann benutzt werden dürften.

Verlag: Südwestdeutscher Verlag für Hochschulschriften Aktiengesellschaft & Co. KG
Dudweiler Landstr. 99, 66123 Saarbrücken, Deutschland
Telefon +49 681 37 20 271-1, Telefax +49 681 37 20 271-0, Email: info@svh-verlag.de
Zugl.: Lausanne, EPFL, Diss., 2004

Herstellung in Deutschland:
Schaltungsdienst Lange o.H.G., Berlin
Books on Demand GmbH, Norderstedt
Reha GmbH, Saarbrücken
Amazon Distribution GmbH, Leipzig
ISBN: 978-3-8381-0524-6

Imprint (only for USA, GB)
Bibliographic information published by the Deutsche Nationalbibliothek: The Deutsche Nationalbibliothek lists this publication in the Deutsche Nationalbibliografie; detailed bibliographic data are available in the Internet at http://dnb.d-nb.de.
Any brand names and product names mentioned in this book are subject to trademark, brand or patent protection and are trademarks or registered trademarks of their respective holders. The use of brand names, product names, common names, trade names, product descriptions etc. even without a particular marking in this works is in no way to be construed to mean that such names may be regarded as unrestricted in respect of trademark and brand protection legislation and could thus be used by anyone.

Publisher:
Südwestdeutscher Verlag für Hochschulschriften Aktiengesellschaft & Co. KG
Dudweiler Landstr. 99, 66123 Saarbrücken, Germany
Phone +49 681 37 20 271-1, Fax +49 681 37 20 271-0, Email: info@svh-verlag.de

Copyright © 2009 by the author and Südwestdeutscher Verlag für Hochschulschriften Aktiengesellschaft & Co. KG and licensors
All rights reserved. Saarbrücken 2009

Printed in the U.S.A.
Printed in the U.K. by (see last page)
ISBN: 978-3-8381-0524-6

Abstract

In the pursuit of a portable gas detector/analyser we studied the components of an ion mobility spectrometer (IMS), which is a device that lends itself well to miniaturisation. The component we focused on was the ionizer. We fabricated a series of micro ionizers with micro electromechanical systems (MEMS) technology, which had a gap spacing between 1 and 50 µm and a thickness from 0.3 to 50 µm. They were used to examine micro discharges as such and as a means of ionization. In our measurements of electrical breakdown in small gaps we confirmed the deviation from Paschen's law for breakdown voltages in gaps below 5 µm. One important result is the identification of conditions for stable DC glow discharge in micro gaps. With planar electrodes we observed stable glow for factors of pressure times gap distance pd up to 0.2 Pa×m in N_2, and up to 0.14 Pa×m in Ar. With thick electrodes the glow range was extended: up to 4 Pa×m in Ar, and 10 Pa×m in laboratory air at atmospheric pressure.

The advantage of using discharges in micro gaps as the ionization principle is the low voltage and power that is necessary to drive a discharge. A prerequisite for using an ionizer in an ion mobility spectrometer is the possibility to operate such an ionizer at high, up to atmospheric, pressure. Our final micro discharge devices were operated in laboratory air for several hours without significant deterioration.

A miniature ion mobility spectrometer was set up, in which miniature and micro discharge ionizers were applied as the ion sources. We extracted ions from micro ionizers and measured mobility spectra of gases and mixtures of gases (air, N_2, Ar). The measured peaks in the mobility spectra varied depending on the gas. In the conclusions we suggest improvements that should increase the resolution and stability of our ion mobility spectrometer, so it may become useful for gas detection. The most important improvements will be a better control of the measurement conditions and of the initial extension of the ion pulse.

In addition to our experimental results we present in this work an overview of the research that has already been done in our area of interest. As the basis of our research, the involved physical theory has been worked out. Consequently, the first chapters contain a compilation of relevant subjects, from the basics of electrostatics to the theory of DC glow discharge as far as we believe it can serve the reader in understanding our results.

Zusammenfassung

Im Bestreben nach einem tragbaren Gas Detektor/Analysator, wurden die Komponenten eines Ionen-Mobilitäts-Spektrometers (IMS) untersucht. Derartige Spektrometer lassen sich vorteilhaft miniaturisieren. Die Komponente, auf die wir uns konzentrierten war der Ionisator. Eine Reihe von Mikroionisatoren wurden mit Methoden der Mikrosystemtechnik hergestellt. Die Elektroden dieser Ionisatoren hatten einen Abstand zwischen 1 und 50 µm und eine Dicke von 0.3 bis 50 µm. Sie wurden zur Untersuchung von Mikroentladungen als solchen und ihrer Verwendung zur Ionisation von Gasen im Besonderen verwendet.

Unsere Messungen elektrischer Durchschläge in kleinen Spalten bestätigten die Abweichung von Paschens Gesetz für Durchschlagsspannungen in Spalten kleiner 5 µm. Ein wichtiges Ergebnis sind unsere Erkenntnisse über die Bedingungen unter denen eine stabile Glimmentladung in Mikrospalten erreicht werden kann. Mit ebenen Elektroden beobachteten wir stabile Glimmentladungen für Faktoren von Druck mal Spaltabstand pd von bis zu 0.2 Pa×m in N_2, und bis zu 0.14 Pa×m in Ar. Mit dicken Elektroden erweiterte sich der Glimmbereich: Bis zu 4 Pa×m in Ar, und 10 Pa×m in Laborluft bei Atmosphärendruck.

Der Vorteil elektrischer Entladungen als Ionisationsprinzip in Mikrospalten liegt in der vergleichsweise geringen Spannung und Leistung, die zu deren Betrieb notwendig ist. Eine Vorraussetzung für die Anwendung eines Ionisators in einem Ionen-Mobilitäts-Spektrometer ist die Möglichkeit, einen solchen Ionisator bei hohem, bis zu atmosphärischem Druck verwenden zu können. Die Mikroionisatoren, zu denen wir schliesslich gelangt sind, wurden in Laborluft während mehrerer Stunden betrieben, ohne einen wesentlichen Schaden davonzutragen.

Ein Miniatur-Ionen-Mobilitäts-Spektrometer wurde aufgebaut, in welchem wir Miniatur- und Mikroionisatoren als Ionenquellen eingesetzt haben. Wir extrahierten Ionen von Mikroionisatoren und maßen Mobilitätsspektren von Gasen und Gasmischungen (Luft, N_2, Ar). Eine Gasabhängigkeit der gemessenen Peaks in den Spektren wurde gezeigt. In unseren Schlussfolgerungen schlagen wir Verbesserungen vor, welche die Auflösung und Stabilität unseres Miniatur-Ionen-Mobilitäts-Spektrometers so weit erhöhen sollten, daß es zur Gasdetektion nützlich wird. Wesentliche Verbesserungen werden von einer besseren Kontrolle der Messbedingungen und der Anfangsausdehnung des Ionenpulses erwartet.

Zusätzlich zu unseren experimentellen Ergebnissen präsentieren wir in dieser Arbeit einen Überblick über Forschung die auf diesem Gebiet bereits durchgeführt wurde. Als Grundlage unserer Untersuchungen erarbeiteten wir uns die notwendige physikalische Theorie. Die ersten Kapitel enthalten entsprechend eine Zusammenstellung relevanter Themen, von den Grundlagen der Elektrostatik bis zur Theorie von Gleichspannungsglimmentladungen, soweit sie dem Leser zum Verständnis unserer Ergebnisse dienen können.

Contents

1 **Introduction** 1
    1.1 The scope of this thesis 1
    1.2 Spectrometric gas analysis 2
    1.3 State of the art 4
        1.3.1 Micro discharges 4
        1.3.2 Miniature ion mobility spectrometers 5
        1.3.3 Micro mass spectrometers 5

2 **Basics of electrostatics, electrodynamics, and gas dynamics** 7
    2.1 Electrostatics 7
        2.1.1 Basic expressions of electrostatics 7
        2.1.2 Calculation of fields 8
        2.1.3 Charges and ions 10
    2.2 Gas dynamics and charged particle kinetics 10
        2.2.1 Movements of electrons and ions in electric fields 11
        2.2.2 The mean free path 12
        2.2.3 Mobility and ion filtering by mobility 14

3 **Ionization of gases** 21
    3.1 Principles of gas ionization 21
    3.2 Electrical discharges in gases and in vacuum 23
        3.2.1 Electric breakdown 23
        3.2.2 Types of DC discharges 38
    3.3 Choice of an ionization method 45
        3.3.1 A comparison of ionization methods 45
        3.3.2 Industrial ion sources 47
        3.3.3 Choice of a method 49

4 **The DC glow discharge** 51
    4.1 Characteristics of DC glow discharges 51
        4.1.1 The cathode layer and the Debye length 52
        4.1.2 Anode layer 55

Contents

    4.2   Corona discharge 56
          4.2.1   Development of a corona discharge..56
          4.2.2   Intermittent corona discharge 58
          4.2.3   Current oscillations in dark discharges...60

5   Micro gas discharge devices 63

    5.1   Experimental setup..63
          5.1.1   Micro electrode designs and fabrication..63
          5.1.2   Setup for micro ionizer experiments..66
    5.2   Discharge experiments with micro electrodes 67
          5.2.1   Breakdown voltage 67
          5.2.2   Glow range..72
          5.2.3   Effects of discharges on electrodes..72
          5.2.4   Special electrode profiles: Rogowski, Bruce and our own......................78
    5.3   Simulation of glow discharge oscillations 79
          5.3.1   Description of an oscillation cycle...80
          5.3.2   Model circuit for oscillating discharges...80
          5.3.3   Micro discharge experiments and comparison with simulations...............84
          5.3.4   Discussion...87
    5.4   Conclusions...89
          5.4.1   The influence of experimental parameters..89
          5.4.2   Conclusions from microionizer experiments..92

6   Ion extraction, filtering and detection 95

    6.1   Miniature ionizer...95
          6.1.1   Glow discharge in a miniature ionizer...95
    6.2   Ion extraction 96
          6.2.1   Ion extraction with and without grid..96
          6.2.2   Grid 97
          6.2.3   Pulsing 97
          6.2.4   Detector and signal amplification 99
          6.2.5   First extraction experiment 99
    6.3   Miniature ion mobility spectrometer (IMS)...101
          6.3.1   The drift chamber..103
          6.3.2   Measurements with the miniature analyser 104
          6.3.3   Microionizers as ion sources...110

	6.4 Conclusions and suggestions	112
7	**Conclusions and outlook**	**115**
	7.1 What has been achieved	115
	7.2 Outlook	116

Acknowledgements ... **117**

References ... **119**

Appendix ... **I**
- A. Signs, units, constants, and abbreviations ... I
  - Abbreviations ... IV
- B. Properties of gases ... IV
- C. List of used instruments ... V
- D. Vacuum system ... VI
- E. Pulse generation circuits ... VII
  - Pulse generator I ... VII
  - Pulse generator II ... VII
- F. Micro ionizer processes ... VIII
  - Type Plan ... VIII
  - Type I-Plan ... IX
  - Type Si-Bulk ... IX

1 Introduction

Gases are everywhere. The air we breathe consists of gases: we can feel them and some gases we can smell. However, they also exist captured in liquids and solid materials, and even, though at a very low density, in outer space. Some gases are vital for life on earth, like oxygen (O_2) and carbon dioxide (CO_2), others are toxic or otherwise dangerous. There are many cases where gases need to be detected: finding leaks, fire alarms, drugs or explosives detection. Here, it is not important to measure the exact concentration of a gas, and relatively simple gas sensors are good enough. In other cases, the exact concentration of a gas in a mixture of gases must be measured: production (process control), food testing, environmental control, medicine. Devices that are more sophisticated are then necessary. Gas chromatography together with spectrometric methods is often applied. Disadvantages of these commonplace highly accurate laboratory devices are their high price, difficult operation and large size. For more specific cases, gas sensors and detectors can be used, that exist in great variety. These are often inexpensive and sometimes highly accurate, but still not available for all desired applications. To overcome these drawbacks was our motivation to study new types of miniaturised gas sensors that are based on ionization of the gases. In these devices the ions are created, then separated and detected over time, creating a sample specific spectrum. The selected principle for ion separation was *ion mobility* because of its relative instrumental simplicity and the viability of miniaturisation. This choice is supported by a remark of Spangler [SPA93]: "It is *easy for anyone* to build an IMS (ion mobility spectrometer) for use in his own laboratory."

1.1 The scope of this thesis

The two main aspects studied in this thesis are

- microfabricated ionizers to be applied in a micro ion mobility spectrometer
- a miniaturized ion mobility spectrometer (IMS)

For the development of micro ionizers, we had to study electrical discharge and ionization phenomena in theory and experimentally, especially in small electrode gaps on the order of 10 µm. We found that very little is known so far about discharges in such small gaps, while discharge at a larger scale has already been the subject of intensive research for more than a century. The reason for using discharges in micro gaps as the ionization principle was the low voltage and power necessary to run a discharge, and the possibility to operate such an ionizer at high, up to atmospheric, pressure. We concentrated on DC glow discharge, and the creation of plasmas in micro systems. The main practical work comprised the design and microfabrication of ionizers, and their assessment in the laboratory.

1 Introduction

The main features we expect from a miniature/micro IMS system are:

- No vacuum needed, operation at or near atmospheric pressure
- No radioactive ionizer (The ionization source in conventional IMS is often a radioactive ^{63}Ni foil)
- Low power consumption
- Hand held, easy to operate
- Essential parts microfabricated
- Low price

One main point was the extraction of ions from the ionizer and their acceleration through the separator towards the detector. Here, the theory is quite well known, and our focus was on the practical aspects.

This dissertation summarizes our investigations of discharges and plasmas on the micro scale, implemented with micro electromechanical systems (MEMS) technology, from concept and theoretical studies to design, realization and testing.

1.2 Spectrometric gas analysis

Mass spectrometry is a commonplace method for high precision gas analysis in laboratories. It works by ionizing a sample gas, filtering the ionized gas molecules according to their mass/charge ratio, and detecting the ions that pass the filter. In ion mobility spectrometry, also called plasma chromatography or gas phase electrophoresis, it is not the mass to charge, but the mobility to charge ratio, according to which ions are selected. The most important difference between these two methods is that the mobility filter chamber is filled with a matrix gas, while the mass filter must be highly evacuated. Also, the resolution of a mobility filter does not depend on the length of the drift chamber as much as it does in a mass filter. These are two major advantages for the application in a portable device.

A general gas analyser that comprises an ionizer typically consists of the components shown in Fig. 1. In this work on a miniature analyser, we do without gas separation before ionization. Like that, the system is much simpler, at the expense of selectivity.

1.2 Spectrometric gas analysis

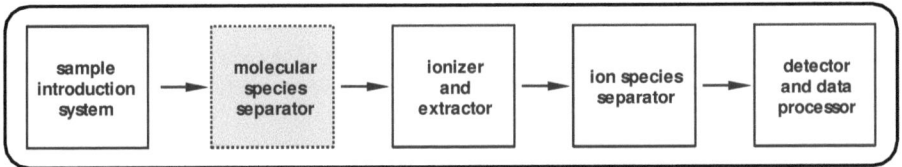

Fig. 1   Typical components of a general gas analyser with ionizer. A molecular species separator (e.g. gas chromatograph) can improve the analysis quality for mixtures of gases but is not always necessary.

A standard ion mobility spectrometer is illustrated in Fig. 2. Fig. 3 shows an example of a spectrum that was measured with a spectrometer of such a kind by Dheandhanoo and Ketkar [DHE03].

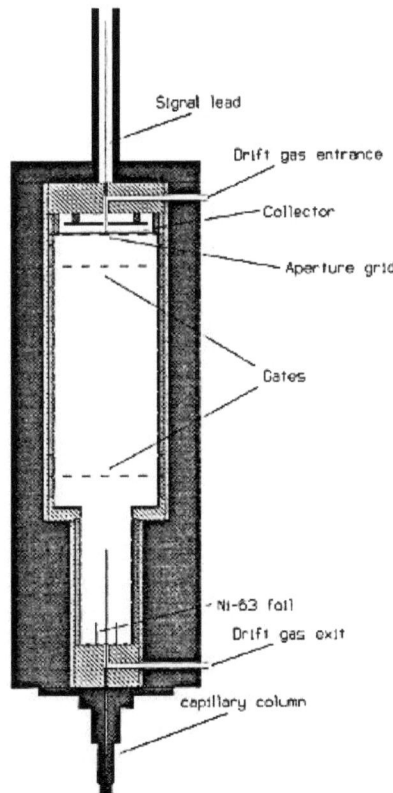

Fig. 2   A standard ion mobility spectrometer with a radioactive ion source [SIW94]. The sample gas enters from the bottom.

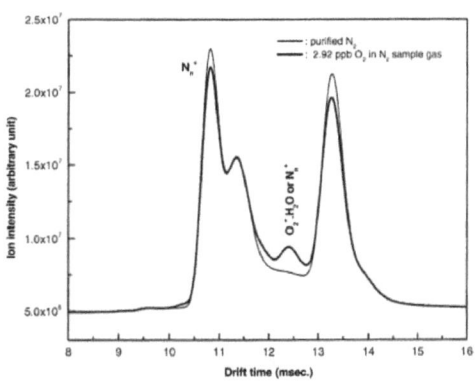

Fig. 3   A sample ion mobility spectrum of N_2 in purified N_2 drift gas. [DHE03]

1.3 State of the art

Here we present examples of the state of the art in the research of and towards micro gas analysers using the principle of gas ionization. A few research groups have been active in this field and a few products are on the market today.

1.3.1 Micro discharges

DC discharges

Dai et al. [DAI99a, DAI99b, DAI99c, DAI99d] constructed a gas sensor that works by measuring the corona discharge current between a needle tip and a metal tube through which the sample gas flows. The current depends on various parameters of the gas, e.g. ionization potential, electronegativity. This sensor can be compared with a thermal detector that measures the heat conductance through a gas. The corona sensor has the advantage of higher sensitivity, but the disadvantage of low stability.

Work very close to our own in micro glow discharges has been done by Eijkel et al. in the development of a molecular emission detector. Their aim was to identify gases by the spectrum of light emission in a glow discharge. When the microstructures did not show sufficient lifetime in DC mode (< 12 h), the group switched to RF plasma, increasing lifetime considerably. Eijkel et al. worked at and slightly above atmospheric pressure. Their etched plasma chamber of 2000×450×200 µm^3 (volume 180 nl) was bordered by 250 nm gold electrodes. The carrier gas was helium. [EIJ00b, EIJ99]

Micro discharges under high pressure from 500 kPa to 15 MPa were examined by Möller et al. in a scanning tunnelling microscope. The electrode distance was varied from 1 to 15 µm. Most stable were discharges in He. The discharge current was less stable in Ar and N_2. The aim of the work was to find a way to use micro discharges for the patterning of photoresist. The resolution achieved was about 1 µm. [MOE00, MOE99a, MOE99b, TER96]

RF discharges

Because of their scientific importance we include RF micro plasmas, although they were not subject of our own research.

Gessner, Scheffler, et al. generated RF discharges between electrodes with distances of 10 to 300 µm, the electrode width was 10 to 400 µm. They used He at pressures up to 150 kPa as the plasma gas. 200 V were applied at 13.56 MHz with a power of 15 to 30 W to get the plasma going. The same electrodes were also used for DC experiments, but at lower pressure: 8 to 20 kPa of He and Ne at 200 to 400 V, 2 mA. A much higher RF power density was necessary to generate the discharge in gases like argon or in molecular gases like N_2, which heated up the electrode systems and therefore decreased their lifetime. Aim of the research is pollution control: the disintegration of toxic waste gases. [GES00, SCP00]

For commercial applications, Jenion, Germany, has been engaged in RF discharges at up to atmospheric pressure in He. The distance of their electrodes was 300 μm. A plasma area of nearly 15×25 mm^2 was achieved. Electrodes made of 500 nm of Cu were connected to a 27.12 MHz generator, delivering a RF power of 60 W. [JEN01, ROT00, SCM01]

1.3.2 Miniature ion mobility spectrometers

Eiceman, a pioneer in ion mobility spectrometry, and Miller, et al. have been working on a miniature radio-frequency mobility analyzer for a while, that has become known under the name FAIMS (High-Field Asymmetric waveform Ion Mobility Spectrometer). The FAIMS technology uses an asymmetric high voltage oscillation applied to a pair of electrodes. This oscillation causes ions introduced into the region between these two electrodes to oscillate, too. Since the waveform is asymmetric, ions will tend more to one of the electrodes. An additional DC voltage keeps ions of a desired mobility on track and allows these ions to pass while others crash into one of the filter electrodes. Instruments applying the FAIMS filter are now sold by the Ionalytics Corporation, Ottawa, Canada. [EIC01a, MIR01, MIR02, MIR99].

A miniature ion filter, not fabricated with microtechnology, has been conceived and its development pursued by Xu et al. [XU00, XU01, XU98, XU99]. Their spectrometer had a drift channel 1.7 mm in diameter and 35 mm in length and the drift chamber comprised 25 stacked copper lenses separated by insulating spacers with mini-resistors. An ultraviolet laser pulse was used to ionize molecules, the initial time spread was 0.76 ms. The main resolution limiting factor was peak broadening due to coulomb repulsion especially in the ionization region. A resolution >10 was achieved with an operating voltage of 500 V.

Another miniaturised ion mobility spectrometer has been presented by Teepe et al. [TEE01]. The microstructured part in their device is the grid, where thin lines have been cut into a "stiff conducting material" by laser ablation.

1.3.3 Micro mass spectrometers

At the TU Harburg, the group of Prof. Müller is working on micro discharges, DC as well as RF [MUE01]. They envisage applying a micro system plasma reactor as an ionization source in a miniaturized mass spectrometer, as well as a variable wavelength light source for an optical micro spectrometer, as a detector at the end of the separation column of a micro gas chromatograph or as a gas detector by itself. A RF micro plasma reactor is planned, where the change in impedance of the RF plasma as well as the emitted optical spectrum are expected to be characteristic for the gas composition.

1 Introduction

Freidhoff et al. have spent some effort on the development of a micro mass spectrometer comprising a miniaturised quadrupole mass filter. Gold-coated optical fibres in V-grooves etched into Si-wafers serve as high precision rods in the filters. [FRE99a, FRE99b, SYM98, TAY98a, TAY98b, TUN98]. More recently Taylor has been working on the project [TAY01, TAY03]. Microsaic Systems Ltd, London, has taken on the commercialization of the spectrometer.

The development of a micro mass spectrometer, ion source and filter, was started some years ago in the group of Müller at the TU Harburg, Germany, and is now pursued at LETI in Grenoble, France by Sillon et al. As a filter, a Wien filter is employed that works with a fixed magnetic field and an electric field that can be adjusted to the ions of interest or be scanned for a spectrum. The ion source has not been integrated on chip yet. Use of an electron impact or a micro plasma source is envisaged. [MUE01, PET00, SIE98, SIE99, SIL01, SIL02].

Yoon et al. integrated an electron impact ion source in the micro time of flight spectrometer, which they currently investigate. [YOO01a, YOO01b]

Telrandhe et al. fabricated cylindrical ion trap microarrays. Ion traps work by trapping ions in a RF field. The stored ions are then released from the trap in a "mass selective instability mode" by ramping the voltage on one electrode, thus obtaining a mass spectrum. The advantage of miniaturization of such filters is the lower voltage required for trapping, the increased mass range of ions that can be trapped, and the higher operating pressure. The trap holes, arranged in an array, had a diameter of 1.5 mm in a wafer of the same thickness. [TEL02]

Whitten et al. also envisage microchip ion traps. So far they have fabricated miniature traps with 12.7 mm trap hole diameters for which they claim to have achieved a mass resolution comparable to or better than for laboratory ion trap mass spectrometers. [KOR00, KOR99, WHI01].

Wiberg et al. published results of work done at the Jet Propulsion Laboratory (JPL) on a miniaturized Gas Chromatograph/Mass Spectrometer (GC/MS) system, comprising miniature system components including turbomolecular pumps, scroll type roughing pump, quadrupole mass filter, gas chromatograph, precision power supply and other electronic components. An array of quadrupole filters was fabricated using the LIGA process. [WIB00, WIB01]

Patents

The following patents have been issued on micromachined mass spectrometers and related microtechnology inventions:

[CHO95, FRE95, FRE96, KOT96, LIN97, MAR98, ROT00, SIT95, SIT96, STA96, WIB01, YOU96]

2 Basics of electrostatics, electrodynamics, and gas dynamics

2.1 Electrostatics

For designing a device for electrical discharge ionization, some understanding of electrostatic fields is a basic prerequisite. Therefore a collection of the involved mathematics and physical theory is given here, providing an overview of electrostatics as far as it is of importance for our design. As it is used here, the expression "electrostatics" involves charges in motion and at rest (under influence of static boundary conditions). Magnetic effects are not covered. As static we regard a voltage that is changing slowly compared with elementary processes in a gas, e.g. during electrical discharge.

2.1.1 Basic expressions of electrostatics

In this section a few basic expressions of electrostatics, used throughout the text are introduced. The units are given in a table in the appendix.

The *voltage* V (also known as the *electromotive force*, EMF) is used for the potential difference between two points. The *electric potential* Φ is the voltage at any point in space with respect to an arbitrary zero potential. Voltage and the *electric field* E both refer to the same physical effect, the force on a charge. The relation between potential and field is:

$$E = -\nabla \Phi \tag{2.1}$$

The electric field strength E is defined as the relation of the force F on a charge to the magnitude of the charge Q:

$$F = QE \tag{2.2}$$

In a parallel plate capacitor, the charges on the plates are distributed uniformly and the field is uniform between the plates. Then the field strength can be calculated from the applied voltage and electrode distance using

$$E = \frac{V}{d} \tag{2.3}$$

The *current* I is defined by the amount of charge Q it brings in a given time t:

$$I = \frac{dQ}{dt} \tag{2.4}$$

The voltage is proportional to the current, as described by Ohm's law:

$$V = RI \tag{2.5}$$

where R is the *resistance*. Especially in highly insulating materials, where the resistance is high, R can become a function of I as well as of V.

In a capacitor, measurements of charge and voltage usually show a linear relation of the form

$$Q = CV \tag{2.6}$$

where C is the *capacitance*. It depends on the geometry of the capacitor (area and plate separation) and the material between the capacitor plates. The material has a relative *permittivity* ε_r (also called the *dielectric constant*) which is a value relative to the permittivity of vacuum ε_0.

$$C = \varepsilon_0 \varepsilon_r \frac{A}{d} \tag{2.7}$$

C can be calculated with the help of the time constant τ that gives the time after which a capacitor is charged from 0 to $1-e^{-1}$ ($\approx 63\%$) of its maximum charge at a given voltage and resistance R, or discharged to e^{-1} ($\approx 37\%$):

$$\tau = C \cdot R \tag{2.8}$$

A capacitor is practically completely charged or discharged after $5 \times \tau$.

2.1.2 Calculation of fields

Laplace and Poisson

Outside of electrodes, electric fields are described by solutions of Poisson's equation:

$$\Delta \Phi = \nabla^2 \Phi = -\rho / \varepsilon_0 \tag{2.9}$$

where Φ is the electric potential and ρ is the charge density.

If there are no charges between the electrodes (*space charge*), then Poisson's equation reduces to the equation of Laplace:

$$\nabla^2 \Phi = 0 \tag{2.10}$$

$$\nabla \Phi = (d\Phi / dx) + (d\Phi / dy) + (d\Phi / dz) = -E \tag{2.11}$$

Boundary conditions are given directly in terms of the applied voltage (see e.g. [SED96]).

Analytic solutions of Laplace's or Poisson's equation can only be obtained for some special and simple geometries. Examples are given below. For more complex geometries, numeric simulations are used to calculate fields.

Examples of Field Distributions in Simple Cases

Analytical Calculations

Spheres and point charges

When a voltage V is applied to a sphere of radius r, a field will appear at the surface of the sphere. This field is equal to a field caused by a point charge at a distance r from the point. Using the relation $Q = CV$, we have:

$$E = \frac{Q}{4\pi\varepsilon_0 r^2} = \frac{CV}{4\pi\varepsilon_0 r^2} = \frac{V}{r} \tag{2.12}$$

since $C = 4\pi\varepsilon_0 r$ for a sphere of radius r.

Coulombs law for two point charges:

$$F = \frac{1}{4\pi\varepsilon_0 \varepsilon_r} \frac{Q_1 Q_2}{d^2} \tag{2.13}$$

where F is the force between the charges Q_1 and Q_2, and d is the distance between them [KUC88].

Tips, wires, and edges

In the space between a parabolic tip with curvature radius r and a plane perpendicular to it at a distance d, the field at a distance x from the tip along the axis is

$$E = \frac{2V}{(r+2x)\ln(2d/r+1)}, \quad E_{max} \approx \frac{2V}{r\ln(2d/r)} \tag{2.14}$$

where V is the voltage between tip and plane [RAI87].

A generalized formula can be given for the field strength at the *surface* of a tip, wire or edge:

$$E = \frac{V}{r} \cdot K \tag{2.15}$$

where K is a geometry factor, which has the form

$$K_{tip} = \frac{2}{\ln(2d/r)}, \quad K_{wire} = \frac{1}{\ln(d/r)}, \quad K_{edge} = \frac{1}{\sqrt{2}} \cdot \frac{1}{\sqrt{d/r}} \tag{2.16}$$

where d is the distance between tip/wire/edge and a plane electrode, and r is the radius of curvature of the tip/wire/edge [BEC71]. The field strength at the wire is only about a factor of two smaller than that at the tip. However, the field strength at the edge is more than one order of magnitude smaller than that at the wire, which is due to the electrostatic field screening of the blade behind the edge. If the same field strength at a given voltage should be obtained with a sharp edge as with a thin wire, the radius of curvature of the edge has to be reduced by about two orders of magnitude. The formulas in (2.16) are applicable only to perfectly smooth surfaces. In reality, no surface is perfectly smooth. Near the surface of any electrode, surface roughness and protrusions dominate the shape of the electric field. For practical discharge calculations, these are taken account of by the so-called *β-factor*. See also §3.2.1 on page 36 for a discussion about field amplification.

2.1.3 Charges and ions

Electrons are negatively charged, protons are positively charged. An atom or molecule carrying an excess of electrons is negatively charged, an excess of protons (loss of electrons) makes it positively charged. In a pair of electrodes, the one that is at a lower potential is called *cathode* and attracts positive ions. The other electrode, carrying less electrons, is called the *anode* and attracts negative ions and electrons. Because positive ions are of much greater interest in this work, we will use the word "ions" for positive ions.

Space charge

The effect of charges on boundaries, for example on insulating side walls is usually examined with Gauss' law: The area integral of the electric field over any closed surface is equal to the net charge enclosed in the surface divided by the permittivity of vacuum,

$$\int \vec{E} \cdot d\vec{A} = \frac{Q}{\varepsilon_0} \qquad (2.17)$$

When the charge which produces the field is not confined to surfaces, the field problem is more difficult. Then the full Poisson equation for the fields must be used. Space charges are most important in plasma and discharge physics, because motions of charges in a plasma or discharge region induce charge rearrangements throughout the entire region. Such rearrangements and their effects are hard to calculate and predict.

At a sharp tip the volume in which the field is very high will be small. In a small volume, relatively few space charges can strongly influence the field. If the anode is the tip, the charges near the anode are the ions that have just been created at the tip and therefore are also positively charged. Like this, they do not neutralize the field, but create a virtual electrode with a greater diameter and therefore a weaker field. In plasmas, the arrangement of space charges near the cathode is even more important, as will be shown in a later section, where the *cathode fall* is introduced (§4.1.1, page 52).

2.2 Gas dynamics and charged particle kinetics

To understand plasmas, it is indispensable to know at least a few basics about gas dynamics and charged particle kinetics, even though we will not try to use this theory for actual plasma calculations here. The practical importance of this theory lies in its application to ion mobility spectrometry. We can learn how an ion, drawn by an electric field, moves through a matrix gas from the plasma electrodes towards the detector. Most of the following equations can be found in the book by J.M. Crowley [CRO86].

2.2.1 Movements of electrons and ions in electric fields

If we assume the movement of only a few charges induced by an external field which is not affected by the motion of the charges, the motion of a charged particle is described by the equation

$$m\frac{d^2\vec{r}}{dt^2} = Q\vec{E}(\vec{r}) + \vec{f} \qquad (2.18)$$

m is the mass, $\vec{r}$ the position of the particle. The electric field $\vec{E}(\vec{r})$ is assumed inhomogeneous, but constant over time. $\vec{f}$ represents non-electrical forces acting on the particle, including magnetic fields and collisions. In vacuum, friction can be neglected, and without magnetic fields (2.18) simplifies to

$$m\frac{d^2\vec{r}}{dt^2} = Q\vec{E}(\vec{r}) \qquad (2.19)$$

Essential for ionization of a molecule by impact of another particle is the energy of the colliding particle. The energy of a charged particle depends on its mass and velocity

$$W = \frac{1}{2}mv^2 \qquad (2.20)$$

In an electric field, the kinetic energy of a charged particle changes when the particle moves from a potential V_1 to V_2:

$$\frac{1}{2}m\left(v_1^2 - v_2^2\right) = Q(V_1 - V_2) \qquad (2.21)$$

The energy difference is therefore proportional to the potential difference between two points. This equation is useful for calculating the energy gain between collisions.

In a homogeneous field, transition time, velocity and energy between two points separated by a distance d in direction of the field can be calculated with the following equations when $W_t(t=0) = 0$:

$$t_t = d \cdot \sqrt{\frac{2m}{QV}}, \quad v_t = \frac{QV}{md} \cdot t_t, \quad W_t = \frac{1}{2}mv_t^2 \qquad (2.22)$$

Using (2.22) and still neglecting collisions, we can estimate the minimum transit time of ions. If we assume a N_2^+ ion, $d = 1$ cm, $V = 800$ V, then

2 Basics of electrostatics, electrodynamics, and gas dynamics

$$t_t = 0.01 \text{ m} \cdot \sqrt{\frac{2 \cdot 2 \cdot 14 \cdot 1.66 \times 10^{-27} \text{ kg}}{1.6 \times 10^{-19} \text{ As} \cdot 800 \text{ V}}} = 0.27 \text{ μs} \qquad (2.23)$$

There are influences besides electric fields which affect charged particles: generation, recombination, convection, diffusion, just to mention the most important. To calculate the transit time of an electron or ion in gas, where collisions are no longer negligible, we need to know about the mean free path and the electron and ion mobility, covered in the next sections.

2.2.2 The mean free path

All of our experiments are taking place in gases, more or less dense, more or less ionized. The gases are assumed to be ideal. These calculations are simplified, and to obtain accurate values the mean free path must be measured. A comprehensive introduction to mean free path calculations can be found in [SED96]. The mean free path is defined as the path length that a particle will travel on average before colliding with another one. It can be calculated with

$$\lambda_m = \frac{1}{\sqrt{2} n \sigma_m} \qquad (2.24)$$

where n is the number of particles per unit volume and σ_m the assumed *effective cross section* of the particles for collisions:

$$\sigma_m = \pi \cdot (2 r_m)^2 \qquad (2.25)$$

r_m is the gas kinetic radius of the particles. r_m is not identical with the physical radius. At low energies of less than about 100 eV it quickly increases, then slowly decreases with higher energies. n in (2.24) is found from the ideal gas law:

$$n = \frac{N_A p}{RT} \qquad (2.26)$$

with Avogadro's number N_A, pressure p, temperature T, and the universal gas constant R.

Since *avalanche* processes, that initiate *electrical breakdown*, depend on electron-molecule collisions, the mean free path λ_e of an electron in a gas is even more important for breakdown and plasma considerations than λ_m. When the electrons are assumed to be much smaller and to move much faster than the other particles, then the following equations apply:

$$\lambda_e = \frac{1}{n \sigma_e}; \quad \sigma_e = \pi \cdot r_m^2 \qquad (2.27)$$

σ_e is the cross section of a particle in the path of an electron. A difficulty for the practical use of the above equations is to find the exact value of the collision cross section. The cross section depends on the applied field. In general, cross sections are measured for the gases and ranges of field strength of interest, but there are also ways to calculate values, see Fig. 4. From $\sigma_e = 2.08 \times 10^{-20}$ m found by Kim et al. [KIM01] for N_2 at an energy of 250 eV, λ_e is found to be about 2 μm.

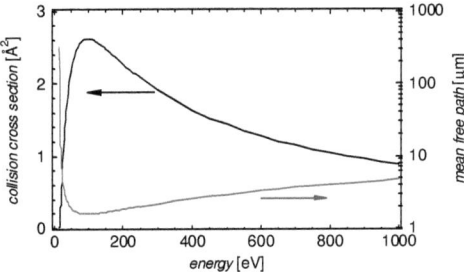

Fig. 4 Ionization cross section of N_2 and mean free path of electrons in N_2 at 300 K and 10^5 Pa. The ionization cross section is smaller than the collision cross section, because peripheral collisions generally do not ionize. [KIM01]

With a known mean free path for an electron, the possibility of a gas discharge in a gap can be estimated. If the gap distance d is shorter than about $2\lambda_e$, no avalanche can occur in the gap. For a glow discharge to develop, d must be much greater than λ_e. In a large enough gap, λ_e can be used to estimate the *breakdown field strength* E_b, if the ionization potential V_i of the gas is known. The energy that an electron gains between two collisions must be greater than V_i:

$$E\lambda_e > V_i \qquad (2.28)$$

Fig. 5 shows an illustration of this equation. For a given gap, the *breakdown voltage* V_b can then be estimated using

$$V_b \approx V_i \cdot \frac{d}{\lambda_e} \qquad (2.29)$$

Fig. 6 shows a graph of breakdown voltages calculated in this simplified way.

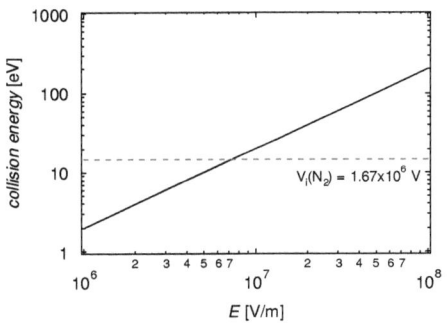

Fig. 5 Electron energy that would be reached by an electron when accelerated by an applied field between collisions. It is assumed that the initial energy and the energy after a collision are 0, and $\lambda_e = 2$ μm. The horizontal line shows the first ionization potential of N_2. Ionization would occur in a field of 7.3×10^6 V/m.

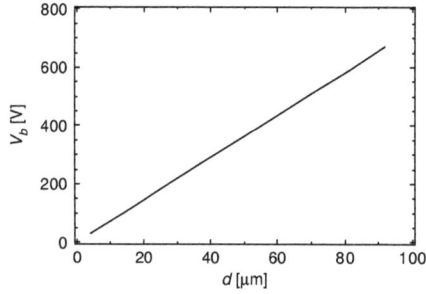

Fig. 6 Breakdown voltage vs. gap width for N_2, with an assumed λ_e of 2 µm.

2.2.3 Mobility and ion filtering by mobility

The mobility μ is defined as the proportionality coefficient between the *drift velocity* v_d of a charged particle and the field [RAI87]:

$$v_d = \mu \cdot E_d \qquad (2.30)$$

The drift velocity v_d is not a strictly linear function of E_d, because the mobility depends on field strength. In an ion mobility spectrometer, E is the constant electric field applied across the drift tube. μ is not constant with changing pressure and temperature in the drift gas. To remove variations due to varying conditions, the reduced mobility μ_0 is introduced:

$$\mu_0 = \mu \cdot \left(\frac{273.15 \text{ K}}{T} \right) \cdot \left(\frac{p}{10^5 \text{ Pa}} \right) \qquad (2.31)$$

Comprehensive collections of data for μ_0 were published by Shumate et al. [SHU86] and by Böhringer et al. [BOE87]. In theory, reduced mobility values are independent of instrumental conditions and accurate drift times can be calculated for any set of instrument parameters. In practice it has been found that μ_0 values do change with instrumental conditions if these conditions are varied over wide ranges.

Mobility of electrons

The mobility of electrons μ_e is

$$\mu_e = \frac{e}{m_e v_m} = \frac{1.76 \times 10^{11}}{v_m} \frac{\text{m}^2}{\text{Vs}^2} \qquad (2.32)$$

where e is the electron charge, and m_e the electron mass [RAI87]. The *collision frequency* v_m is proportional to the pressure p of the gas. Values are found experimentally. Raizer gives a value of 0.03 $\text{s}^{-1}\text{Pa}^{-1}$ for v_m/p for N_2, for example.

2.2 Gas dynamics and charged particle kinetics

For $p = 10^5$ Pa:

$$v_m = 0.03 \text{ s}^{-1}\text{Pa}^{-1} \times 10^5 \text{ Pa} = 3 \times 10^3 \text{ s}^{-1}$$

$$\mu_e = \frac{1.76 \times 10^{11}}{3 \times 10^3} \frac{\text{m}^2}{\text{Vs}} = 5.9 \times 10^7 \frac{\text{m}^2}{\text{Vs}}$$

If the frequency is constant, $\mu_e \propto p^{-1}$ and $v_d \propto E/p$. The energy spectrum and the mean electron energy also depend on E and p not independently, but in the combination E/p. Hence, the drift velocity is a function of the ratio E/p. This *similarity* serves to reduce the amount of measurements and the results can be plotted not as functions of two variables (say, E and p), but as functions of E/p, for example, like $v_d = v_d(E/p)$. It must be mentioned here that according to Osmokrovic [OSM94] similarity laws do not hold for vacuum breakdown. The drift velocity increases with E/p but this growth is not necessarily close to direct proportionality, since v_m and v_d depend on the electron energy distribution [RAI87].

Mobility of ions

For atomic ions, the mobility also depends on the electronic state [CLE97]. For a large polyatomic ion, average collision cross-section determines the mobility. If charge transfer occurs, it considerably reduces the mobility of ions in their own gas [RAI87]. In weak and moderate fields, the ion mobility is given by

$$\mu_i = \frac{e}{m'v_m}; \quad m' = \frac{m_i \cdot m}{m_i + m} \quad (2.33)$$

where m' is the *reduced* mass, m_i is the ion mass, and m the molecular mass. In a field typical of glow discharge, $E/p = 1.3$ V/(Pa×m), we have $v_{id} = 50$ m/s. The corresponding thermal velocity at 300 K is about 400 m/s.

In strong fields, v_{id} is no longer proportional to E/p as it is in weak fields, but to $(E/p)^{1/2}$ [RAI87]. The transition from "moderate" to "strong" field is gradual and begins usually in fields in which ion energies reach about 1 eV, the polarization forces are replaced with short-range forces, and the cross section becomes gas-kinetic. Experimental results for drift velocities have been collected and published for example by Brown [BRO59].

Ions in an ion mobility spectrometer can associate with other neutral species (such as water, nitrogen, etc.) to form ion clusters. The clustered adducts affect not only the mobility characteristics for the ions, but also the diffusion and energy characteristics of the ions. The drift velocity v_d for ion clusters is given by a weighted sum of the drift velocities v_{di} for the individual rapidly exchanging ion species contained within the cluster [SPA99]:

$$v_d = \sum_i x_i v_{di} \quad (2.34)$$

2 Basics of electrostatics, electrodynamics, and gas dynamics

Ion mobility filtering

The separation of ions in the drift chamber of an IMS depends on the different mobility of different ion species, which lead to different drift times in the chamber. The drift time t_d for a drift length l_d is

$$t_d = \frac{l_d}{\mu \cdot E_d} = \frac{l_d^2}{\mu \cdot V_d} \tag{2.35}$$

Pulse Broadening

When ions are inserted into the drift chamber, they are inserted in the form of a pulse. This ion pulse is created by either pulsing the ion source itself, or by continuously ionizing the sample gas and pulsing a grid through which the ions enter the chamber. In either case the pulse has an initial width t_g depending on the pulse length and the extension of the grid or ionizer. Factors that determine the broadening include

- the initial ion pulse width t_g
- broadening by Coulomb repulsion between ions in both the ionization and drift regions: t_C
- spatial broadening by diffusion of the ion packet t_D
- ion-molecule reactions in the drift region

If ion-molecule reactions are neglected, the square of the full width at half maximum is determined by

$$w_{1/2}^2 = t_g^2 + t_C^2 + t_D^2 \tag{2.36}$$

The shape of the pulse is assumed to be Gaussian. An equation is derived as given by Siems et al. [SIW94] to calculate the peak width at half maximum $w_{1/2}$ of the pulse after a drift time t_d:

$$w_{1/2}^2 = t_g^2 + \frac{16 \ln 2 \cdot D \cdot t_d}{v_d^2} \tag{2.37}$$

Coulomb broadening is ignored in this equation. D is the effective diffusion coefficient, which is—like the drift velocity—specific to the ion species:

$$D = kT\mu/q \tag{2.38}$$

therefore

$$w_{1/2}^2 = t_g^2 + \frac{16 \ln 2 \; k}{q} \cdot \frac{Tt_d^2}{V_d} = t_g^2 + 9.57 \times 10^{-4} \cdot \frac{Tt_d^2}{V_d} \tag{2.39}$$

The peaks having a larger delay are broader than the peaks having a smaller delay. The initial ion pulse width is the limiting factor for the peak resolution, because it is the minimum width the peak can attain. If two isomers have mobilities that differ by less than the ion packet expands, they will not be resolved.

To achieve a better fit of experimental results to calculated values of $w_{1/2}$, Siems et al. [SIW94] suggest the following equation for peak width calculations:

$$w_{1/2}^2 = \gamma + \beta t_g^2 + \alpha \frac{T t_d^2}{V_d} \tag{2.40}$$

where α, β, and γ are fit parameters that are specific to the instrument. Equation (2.39) is then valid for an ideal instrument where $\alpha = 9.57 \times 10^{-4}$, $\beta = 1$, $\gamma = 0$.

Resolution

With *resolution*, we denote the capability to resolve two consecutive peaks in a spectrum. The resolution depends on the width of the peaks: the narrower the peaks, the better they can be distinguished. In spectrometry, a great variety of definitions of resolution can be found depending on the application and the choice of the author. We use the following 50% peak-height definition of resolution [SIW94]:

$$R = \frac{t_d}{w_{1/2}} \tag{2.41}$$

This definition of R is easily accessible and has the advantage that one single peak is sufficient for its calculation. When t_g is neglected in eq. (2.39), an approximate equation for the resolving power is found:

$$\frac{t_d}{w_{1/2}} = 32.3 \sqrt{\frac{V_d}{T}} = 32.3 \sqrt{\frac{E_d l_d}{T}} \tag{2.42}$$

From this expression it is apparent that in order to increase the resolving power it is necessary to lower the temperature, increase the drift field or increase the length of the drift tube. If the drift field is increased, there must be a corresponding increase in the buffer gas pressure in order to keep the mobilities in the low-field limit (low-field limit: mobility independent of the drift field). The length of the drift tube is still limited by the expansion of the ion packet by diffusion as it travels through the drift tube. With a buffer gas pressure of tens of kPa, drift fields of several hundred volts per centimetre can be employed and the drift tube can be up to several meters long. The resolving power of such a high resolution configuration is around 200-400 [CLE97].

Elaborating more on eq. (2.40), an equation is found for the resolution that is divided into a part governed by diffusion (R_d), and one that depends on the initial pulse width (R_p):

$$R^{-2} = \frac{\gamma + \beta t_g^2}{t_d^2} + \frac{\alpha T}{V_d} = R_p^{-2} + R_d^{-2} \qquad (2.43)$$

with (2.35):
$$R_p = \frac{l_d^2}{\left(\gamma + \beta t_g^2\right)^{1/2} \mu V_d} \qquad (2.44)$$

$$R_d = \left(V_d / \alpha T\right)^{1/2} \qquad (2.45)$$

Once the initial pulse width is known, these equations can be used to find scaling laws for an IMS tube.

The initial pulse width

The initial pulse width t_g depends on the geometry (extension of the ionizer l_{iz}) and the pulse time. When a grid is used, t_g corresponds approximately to the gate opening time. When the ionizer is pulsed, we assume that the pulse width will at least equal the extension of the discharge divided by the drift velocity:

$$t_g \approx \frac{l_{iz}}{v_d} = \frac{l_{iz}}{\mu \cdot E} \qquad (2.46)$$

If the ionization pulse time t_{iz} is longer than l_{iz}/v_d, this approximation is invalid. As a simplification one can then use the applied ionization pulse t_{iz} as the pulse length t_g. An additional (Coulomb) broadening of the initial pulse is caused by the high ion density in the discharge itself [XU00].

Scaling of a drift tube

The scaling of l_d and V_d as calculated with (2.43) is illustrated in Fig. 7. We see that the resolution theoretically increases with the drift length up to a maximum determined by diffusion, and has a maximum at a voltage determined by diffusion as well as initial pulse width. What is not taken into account here, is the dependency of the mobility coefficient on field strength.

The maximum resolution increases for longer tubes and higher voltage. However, the signal strength reduces with tube length, and for long tubes (~ 1 m), the ideal voltage becomes impractically high (> 10 kV). In the design of a drift tube, a compromise must therefore be found between signal strength, applied drift voltage and tube length. The designer must also have the actual mobility and boiling temperature of the molecules in mind that he wishes to detect.

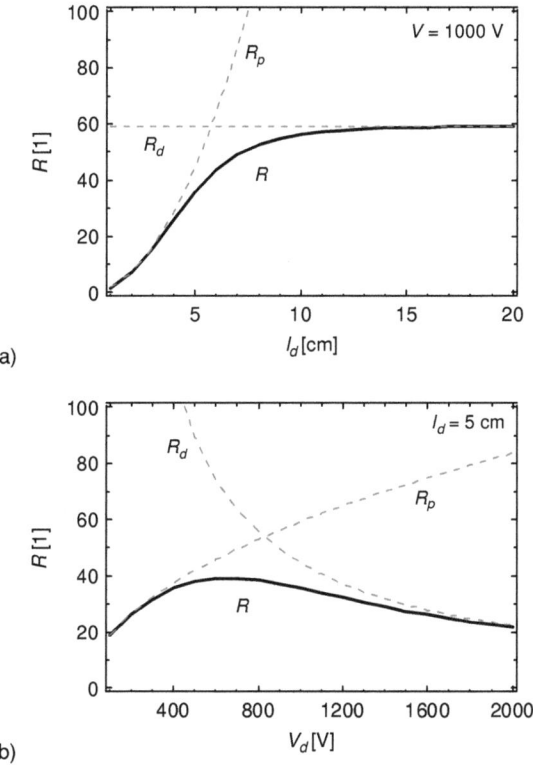

Fig. 7 Scaling of l_d and V_d in an ion mobility drift tube. The resolution is plotted over the drift length l_d while a) the drift voltage b) the drift length is kept constant. R_d is the diffusion-only resolving power, R_p is the pulse-width-only resolving power. $T = 25\ °C$, $\mu = 2.8\times10^{-4}\ m^2/Vs$, $t_g = 0.2\ ms$.
a) V_d is kept constant at 1000 V while l_d is varied from 0 to 20 cm. As a result, the resolution R tends towards a maximum determined by diffusion.
b) l_d is kept constant at 5 cm while V_d is varied from 0 to 2000 V. As a result, we find a maximum for the resolution at a voltage of 660 V.

Numerical examples

To give numerical examples, typical values for a miniature IMS are applied:

$T = 25\ °C$, $\mu = 2.8\times10^{-4}\ m^2/Vs$ (NO^+ in N_2)

$l_d = 2\ cm$, $l_{iz} = 200\ \mu m$, $V = 500\ V$

We find $D = 7.3\times10^{-6}\ m^2/s$, $E = 2.5\times10^4\ V/m$, $v_d = 7\ m/s$, $t_g = 28.6\ \mu s$, $t_d = 2.9\ ms$.

The pulse width $w_{1/2}$ at the end of the drift chamber is then 74 µs. For the resolution we find $R \approx 39$.

With the set of equations collected in this section we have a tool for the assessment and design of our own measurements and devices. For example, we can find the optimum voltage to apply to a drift chamber when all other parameters are given. Keeping t_g at 28.6 µs and optimizing the above example numerically, we find a maximum R of 41 with a drift voltage of 700 V, i.e. $E = 35$ kV/m. If the drift length was reduced to 1 cm, 280 V ($E = 28$ kV/m) should be applied to achieve $R = 26$.

3 Ionization of gases

In this chapter we treat the theory of the ionization of gases. We first give an overview of common ionization methods. For our ionizer we wish to use electric discharge ionization as the ionization principle. Discharges are therefore given special attention in the ensuing sections. Our theoretical reflections will then lead to the choice of the discharge that we pursue thereafter: the DC glow discharge.

3.1 Principles of gas ionization

A gas consists of atoms and/or molecules, which are electrically neutral, because the sum of positive charges in their nuclei (protons) and the sum of negative charges in their shells (electrons) equals zero. In nature, gases are likely to contain some ions as well. These are atoms or molecules ionized by cosmic rays, for example. The density of these "natural" ions is usually negligible. Only when a significant part of a gas is ionized we speak of a *plasma*. Atoms are maintained as a stable unit by the potential gradient between nucleus and electrons. This gradient can be calculated from the principles of quantum mechanics and at the outer boundary of most atoms it is about 10^{10} V/m. To ionize a neutral particle, an electron must be extracted from the electron cloud, or added to it. Both mechanisms require energy.

When A and B are neutral particles, AB a molecule, e an electron, and "-" denotes negative and "+" positive charging, the possible kinds of ionization are listed in Table 1.

Table 1   Possible kinds of ionization of a neutral molecule AB. [BUD92]

AB	$\rightarrow$	$AB^+ + e^-$	(1)
AB	$\rightarrow$	$AB^{2+} + 2e^-$	(2)
AB + e^-	$\rightarrow$	AB^-	(3)
AB	$\rightarrow$	$A^+ + B^-$	(4)
AB	$\rightarrow$	$A^+ + B + e^-$	(5)
AB + e^-	$\rightarrow$	$A^- + B$	(6)

These kinds of ionization are enabled by various methods the most common of which are briefly introduced. The numbers in parentheses are references to possible respective reactions in Table 1.

3 Ionization of gases

The mentioned and other methods are discussed in more detail in the following chapter, where eventually the method of our choice (glow discharge) is selected.

Field ionization

High electric fields pull electrons out of the particles' electron cloud, thus creating positively charged ions. (1, 2, 5)

Impact ionization

Collision with high energy electrons or particles can force electrons out of the shell, excite particles, or break molecules. Excited molecules may pass their energy to others thus ionizing them. (1-6)

Thermal ionization

If enough energy is supplied to the system by heating, the binding energy of atoms and molecules is overcome and the particles are ionized. *Flame ionization* is a combination of thermal and chemical effects. (1-6)

Radioactive ionization

Radioactive irradiation excites particles and ionizes neutrals. Natural ionization by cosmic rays is an example. (1, 2, 4, 5)

According to Tabrizchi et al. [TAB00], a radioactive ^{63}Ni foil is the usual ionization source of conventional IMS. This radioactive source emits β particles with an average energy of 67 keV. Other radioactive materials have also been used. The benefits of such ionization sources are their simplicity, stability, and noise-free operation. Further, there is no need for an extra power source for ionization. The problems are the regular leak test and special safety regulations that are required to work with such materials. Licensing and waste disposal add to the limited acceptance of IMS on the market. The rate of ion generation with the ^{63}Ni source is not high: the signals, at most, are on the order of 10^{-11} A.

Photoionization

The principal mechanism for photoionization of an atom or molecule is photon absorption and electron ejection to form an ion. The photon energy is typically just above the ionization potential. Usually a strong UV light source provides the photons for ionization. (3)

3.2 Electrical discharges in gases and in vacuum

Processes where ionized particles play an important role in the electric charge transport are called *discharges*. There are many kinds of electrical discharges. In this section, an overview of discharge types is given and the ones relevant for this work are treated in more detail. We begin where discharge starts: breakdown.

3.2.1 Electric breakdown

To give a definition, in the most general sense, *electric breakdown* is the process of transformation of a nonconducting material into a conductor as a result of applying to it a sufficiently strong field. This occurs when applying a voltage at least equal to the so-called *breakdown voltage* or *ignition potential*. The observable effects are breakdown of the interelectrode voltage, and—in most cases—the emission of light visible to the naked eye. In gases, in a homogeneous field, breakdown initiates a discharge that will be *self-sustained* and continue as long as the applied field is high enough. In inhomogeneous fields, *corona discharges* may be initiated before breakdown occurs. We speak of breakdown only once ionized gas spans the electrode gap. [RAI87]

A current is called self-sustained when it flows even in the absence of an outside source of electrons. The processes of secondary emission and multiplication become self sustaining if the ions from multiplication between the electrodes release sufficient secondary electrons from the cathode to replenish the population of ions in the gap [BRA00]. Table 2 shows a list of mechanisms by which electrons or the energy of electrons may get lost.

Table 2   List of electron loss and electron energy loss mechanisms. [RAI87]

Mechanism	Conditions/Comments	Rate of electron loss/energy loss
Excitation of electron states of atoms and molecules	energy loss	~ 0-100%
Excitation of molecular vibration and rotation	energy loss	~ 0-100%
Elastic collisions	energy loss	~ 0-100%
Electron precipitation on the walls (by diffusion)	loss	linear to electron density
Attachment in electronegative gases	loss, formation of negative ions	linear to electron density. [STR99a] In air, the main electron-loss process is electron attachment to oxygen.
Recombination	loss	<< below breakdown, proportional to the electron density squared
Transport to the anode	loss	>>, linear to electron density

3 Ionization of gases

Conduction Mechanisms in Electrode Gaps

Before breakdown, a number of factors can contribute to current between electrodes, see Fig. 8. These factors depend on the materials, the geometry and the atmosphere (gases/vapours, pressure, temperature) around the electrodes. After breakdown, electron transport through the ionized gas between the electrodes is the main contribution to the current that increases by many orders of magnitude when breakdown occurs. An overview of current transport mechanisms is given in Table 3. Each of these factors can help initiate breakdown. This is one reason, why the actual breakdown voltage is hard to predict.

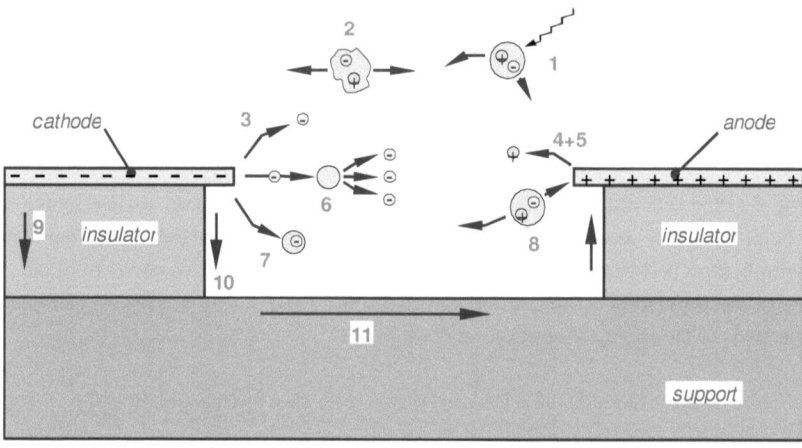

Fig. 8 Illustration of conduction mechanisms in an example of a high voltage/high field electrode gap before breakdown.

1: ionization by radiation (photons, cosmic rays)
2: charge transfer by particles
3: e⁻-emission (field, secondary, photo)
4: field desorption and ionization
5: impact ionization of electrode surface materials
6: charge multiplication in avalanches (by impact)
7: evaporation of charged adsorbed particles
8: dissociation of neutral particles
9: conduction through insulator and polarization
10: conduction over adsorbed layers
11: conduction through support

Table 3 List of mechanisms that contribute to the total current between electrodes.

Mechanism	Conditions/Comments	Current contribution (I_c)
Cosmic rays (1)	Low rate, important before Townsend discharge. $\sim 10^9$ m^{-3}s^{-1} [VEL01], $\sim 10^7$ m^{-3}s^{-1} [RAI87]	<< pA
Photoionization (1)	May be important in coronas	[RAI87]
Electromagnetic waves (1)	Microwaves, laser	Rate of ionization 0-100% [RAI87]
Charge transfer by particles (2)	Field strong enough to pull particle off an electrode surface. Such particles may trigger a breakdown.	Single peaks (micro discharges) $I \sim \mu A$ in microgaps
Field emission (3)	High field strength	pA in μ-gap before breakdown (Fowler-Nordheim equation)
Secondary electron emission (3)	On high energy impact of ions, see also [RAI87]	The number of secondary electrons released by impact of a 10 kV ion is of the order of 4, depending on the kind of the ion and the cathode. [BEC71]
Secondary photo-processes (3)	Important in positive coronas	See [RAI87]
Field desorption (4)	See field ionization	=> Charge transfer by particles
Field ionization (4)	High field strength	Total tip FI currents to be measured are of the order 10^{-10}-10^{-8} A [BEC71]
Impact ionization of electrode surface materials (5)	Strong enough field, initiating electrons exist. Field not disturbed by space charge.	Comparable to ionization in gas (6)
Avalanche (6)	Strong enough field, initiating electrons exist. Field not disturbed by space charge. This is the most important process during breakdown.	Homogeneous field [RAI87]: $I = I_0 \exp(\alpha d)$, α: Townsend's coefficient Inhomogeneous field [SPY95]: $I = I_0 \exp \int_{h_1}^{h_2} \alpha(h) \, dh$, h: axial position
Evaporation of electrode material (7)	High emission current density leads to pressure increase, likely to breakdown. Most important in vacuum breakdown.	=> Charge transfer by particles, but continuous

Mechanism	Conditions/Comments	Current contribution (I_c)
Dissociation (8)	Molecules that easily dissociate, in high field near/on both electrodes (catalysis depending on electrode material).	[MAD93]
Conduction through insulator (9)	Low if intact, defects possible.	$I_{ins} = V_{ins}/R_{ins}$: glass: $\rho = 10^{13}$ to $>10^{15}$ Ωm polymers: $\rho = 2\times10^7$ to $>10^{15}$ Ωm semicond.: $\rho = 10^{-5}$ to 10^6 Ωm
Conduction over adsorbed layers (10)	Esp. water layers, which can be evaporated.	$I_{ads} = V_{ads}/R_{ads}$
Conduction through support (11)	If conductor, then high and linear, otherwise negligible.	$I_S = V_S/R_S$
Conduction through gas	After breakdown: depending on charge (ion/electron) density/degree of ionization.	Conductivity $\kappa = e\times\mu\times n$ (e: electron charge, μ: mobility, n: charge density) [WAS72a]

Conduction in ionized gases

Under the force of an electrical field E, charges of density n form a directed charge flow that is described by

$$j = e \cdot \mu \cdot n \cdot E = \kappa \cdot E \tag{3.1}$$

where

$$\kappa = e \cdot \mu \cdot n \tag{3.2}$$

is the electrical conductivity. $\rho = 1/\kappa$ is called the specific resistance. The mobility of charges μ depends on the gas type, pressure, velocity distribution and the degree of ionization. Because the mobilities of massive ions are hundreds of times smaller than those of electrons the electrical current is practically exclusively due to electrons, except in rare cases when the ion densities n_+, n_- exceeds by a significant number of times the electron density n_e. In low pressure discharges electrons may have an average energy of only about 1 eV. The ions on the other hand can be regarded as immobile most of the time [RAI87, WAS72a].

Examples for specific resistance and current density:

Hg-Plasma: 6.2×10^{-1} Ωcm; 1.6×10^3 A/m^2
Copper (bulk metal): 1.8×10^{-6} Ωcm; 5.6×10^6 A/m^2

The conductivity of a weakly ionized gas is mostly determined by the degree of ionization n_e/N. In strongly ionized plasma, the scattering of electrons by ions impedes their drift along the field as much as that by molecules. As a result of a high Coulomb cross section, electron-ion collisions are already appreciable at ionizations greater than 10^{-3}. [RAI87]

The initiation of electrical breakdown

Because it is an electrical breakdown that initiates a plasma, we begin by describing how breakdown itself is initiated. Like this, we are able to predict to a certain degree what voltage is necessary for discharge initiation under given conditions, or how a discharge gap must be designed to operate at a given voltage.

Electron Avalanche

The avalanche has already been mentioned several times. It denotes a process of multiplication of electrons in a series of impact ionizations. If one applies a high voltage to the electrodes in an atmospheric pressure gas, free electrons will accelerate and collide with molecules of the gas. Hereby more electrons are released, which in turn multiply, thus creating an avalanche. A free electron must be present to initiate the avalanche. It can originate from detachment of a negative ion or from ionization through cosmic rays, for example. In an avalanche, the current of electrons leaving the cathode is enhanced by a factor $\exp(\alpha d)$, where α is the Townsend coefficient for ionization:

$$I = I_0 \cdot e^{\alpha d} \tag{3.3}$$

Secondary Emission of Electrons

While field electron emission is decisive in breakdown, the emission of secondary electrons from the cathode is important in sustaining a discharge. The processes of secondary emission and multiplication and therefore the discharge is *self sustained* when current flows even in the absence of an outside source of electrons. This is the case when the ions from multiplication between the electrodes release sufficient secondary electrons from the cathode upon impact to at least replenish the population of ions in the gap. [BRA00]

With secondary emission taken into account, the steady discharge current is given by

$$I = \frac{I_0 \cdot e^{\alpha d}}{1 - \gamma(e^{\alpha d} - 1)} \tag{3.4}$$

The emission I_0 is caused by positive ions, photons, and metastable atoms produced in the gas as a result of ionization and excitation of atoms by electrons. $(e^{\alpha d}-1)$ is the factor by which the original electron current is increased due to ionization of gas molecules, where α is the ionization coefficient. γ is the effective secondary emission coefficient for the cathode: One electron emitted by the

3 Ionization of gases

cathode produces ($e^{\alpha d}$-1) ions that, hitting the cathode, knock out γ electrons each (if this is the ion-electron emission). As long as the denominator of (3.4) is greater than zero, the current remains non-self-sustained. The condition for initiating a self-sustaining discharge found by Townsend in 1902, also called the transition condition, is

$$\alpha d = \ln\left(\frac{1}{\gamma}+1\right) \tag{3.5}$$

As a result of secondary emission, the region of linear growth of ln(I) with d turns steeply upward. This process occurs when the denominator of (3.4), which is very close to unity at small enhancement coefficients αd, tends to zero as αd increases. When the denominator becomes zero, breakdown takes place, a self-sustaining discharge is formed, and the expression (3.4) becomes meaningless. If $\gamma \sim 10^{-1}$–10^{-3}, an electron triggers a self-sustained discharge if it takes part in $\alpha d/\ln 2 \approx \ln \gamma^{-1}/\ln 2$ (3 to 10) ionizing collisions along the path d [RAI87].

Secondary emission depends on the cathode material: Metal oxides on metal show a high gain, some > 30 for electron impact, while metals have a gain below 2. The gain of positive ions is much lower than that for electrons, since in an ion-valence-electron collision only a small fraction of the ion's energy is transferred to the electron, a consequence of the large mass difference. Thus, the bombarding ion energy must be much greater than that of electrons to give a comparable secondary emission gain. Still, ion-induced secondary emission plays an important role in gas discharges. [SED96]

The Paschen curve

In 1889, F. Paschen published a paper [PAS1889] which set out what has become known as Paschen's Law. The law essentially states that the breakdown characteristics of a gap are a function of the product of the gas pressure p and the gap length d. The product pd is a measure of the number of collisions an electron makes by crossing the gap. The pressure should actually be replaced by the gas density, which is affected by the temperature as well as the pressure of the gas. For air, and gaps on the order of a millimetre, the breakdown is roughly a linear function of the gap length: $V = 0.03\ pd + 1.35$ kV. [LUX01]

Many other factors have an effect on the breakdown of a gap, such as radiation, particles (dust), electrode shape and surface irregularities. Paschen's law reflects the Townsend breakdown mechanism in gases, that is, a cascading of secondary electrons emitted by collisions in the gap. Typically, the Townsend mechanism (and by extension Paschen's law) apply at pd products less than 1.3 kPa×m, or gaps around a centimetre at one atmosphere. Furthermore, modifications are necessary for highly electronegative gases because they recombine the secondary electrons very quickly. In general, an equation for breakdown is derived, and suitable parameters chosen by fitting to empirical data. The Paschen law for parallel plates:

Breakdown voltage: $V_b = B \dfrac{pd}{C + \ln(pd)}$ (3.6)

where: $C = \ln\left(\dfrac{A}{\ln(1+1/\gamma)}\right)$

Breakdown field strength:

$$E_b = \dfrac{V_b}{d}$$ (3.7)

The minimum is given by:

$$(V_b)_{min} = B(pd)_{min}$$ (3.8)

and: $(pd)_{min} = \dfrac{\overline{e}}{A}\ln(1+1/\gamma)$ (3.9)

$\overline{e}$ is the natural logarithm number, A and B are gas dependent coefficients:

For air:
$A = 11.5\ \mathrm{m^{-1}Pa^{-1}}$
$B = 279.6\ \mathrm{Vm^{-1}Pa^{-1}}$
and $\gamma = 10^{-2}$
=> $C = 0.9$

For N_2:
$A = 9.2\ \mathrm{m^{-1}Pa^{-1}}$
$B = 214.2\ \mathrm{Vm^{-1}Pa^{-1}}$
and $\gamma = 2 \times 10^{-2}$
=> $C = 0.85$

Townsend's second ionization coefficient γ is covered in more detail in the next section. It increases with distance from pd_{min}. V_b decreases when γ increases. This model is only valid for low currents where the electric field is not significantly disturbed by space charge ($\leq 1\ \mu A$). Since the breakdown depends on *field strength* for small *pd* factors, the curve found by Paschen is different for inhomogeneous fields: The minimum breakdown potential and the corresponding pd_{min} values increase with the inhomogeneity of the field [HEL97]. Examples for Paschen and Paschen-like curves are shown in Fig. 9.

3 Ionization of gases

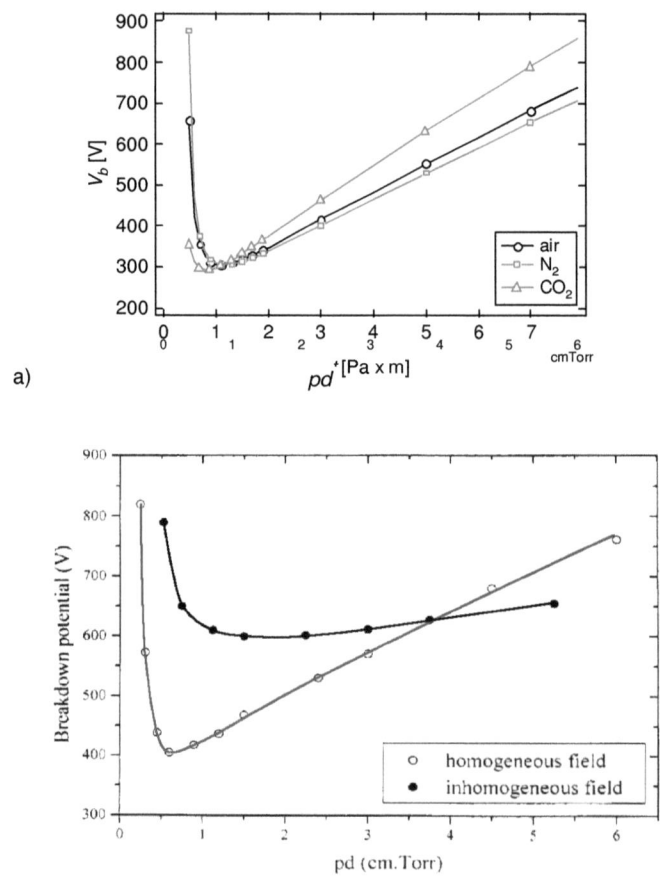

Fig. 9 Paschen curves for various gases. a) Curves calculated with formula given by Raizer [RAI87]. b) Paschen (-like) curves as measured by Held et al. [HEL97] for homogeneous and inhomogeneous fields in air.

Table 4 shows a list of example data for breakdown in various gases found by Naidu [NAI99].

Table 4   Minimum sparking potentials for various gases in homogeneous fields [NAI99]. d is calculated with $d = pd/p$.

Gas	V_s min (V)	pd at V_s min (Pa×m)	d at $p = 10^2$ kPa (µm)
Air	327	0.76	7.6
Ar	137	1.20	12.0
He	156	5.33	53.3
CO_2	420	0.68	6.8
N_2	251	0.89	8.9
O_2	450	0.93	9.3

Breakdown in small gaps or at low pressures

The limit of Paschen's law

It has been shown that Paschen's law is not valid for gaps of less than several µm. For Paschen's curve it does not matter if the pressure or the electrode distance is changed. In a small gap, when surface roughness becomes non-negligible, changing the pressure must have a different influence than changing the distance. Several groups examined electrical breakdown in such gaps [ONO00, TOR99b]. Below 4-5 µm between electrodes, as for example in microactuators and motors, discharge is not well understood. It is stated that at these low separations the gross electrode shape had little, but surface quality had great influence. This is not surprising, since the actual distance is hard to measure exactly and has a great influence. Even "micro"-particles are large in this scale and any protrusions that exist on the surface or are formed or attracted during operation (hillocks, dust) will have a strong influence on the emission and breakdown characteristics of the gap. Some authors reported that the breakdown threshold differs from estimated values from Paschen's law when gaps are below 4-5 µm even for smooth electrodes [GER59, TOR99a]. In our own measurements we also found a significant deviation from Paschen's law below 5 µm.

At large values of *pd*, breakdown is initiated by a *streamer*. Streamers are explained in section 3.2.2 on page 43. Under small *pd* values the streamer breakdown mechanism is replaced by the Townsend breakdown mechanism. Further decrease of *pd* value results in vacuum breakdown. The *pd* limits for these mechanisms are not very sharp. Osmokrovic [OSM94] gave an overview of the relationship between the value of *pd* and breakdown mechanisms which is presented in Table 5.

3 Ionization of gases

Table 5 Comparison of breakdown mechanisms according to Osmokrovic [OSM94].

Breakdown mechanism	pd value	Characteristics
Streamer	large	Secondary processes in gas dominate those on electrodes. Calculation of breakdown voltage: $$\int_0^d (\alpha-\eta)\,dx = 10.5 \qquad (3.10)$$
Townsend	medium	Secondary processes on electrodes dominate those in gas. Calculation of breakdown voltage: $$\gamma \int_0^d \exp\left(\int_0^x (\alpha-\eta)\right)\alpha\,dx = 1 \qquad (3.11)$$
Vacuum	small	Interelectrode gap smaller than the mean free path of electrons. Breakdown due to electrode material evaporation.

η in Table 5 is the coefficient for the number of electrons per cm of distance in direction of the field attached to electrically negative atoms or molecules. The terms "large", "medium" and "small" for values of pd can be related to the Paschen or a Paschen-like curve. "large" then means on the right side of the curve, "medium" near the minimum, and "small" is to the left side of the minimum. The coefficients α, η, and γ have no constant values for any particular gas or gas mixture but vary with pressure and electric field.

α can be roughly calculated with $\alpha \approx (\ln 2)/\lambda$. An electron passing through a potential difference of 1 V generates α/E electrons (pairs of ions). In order to create one pair, it must be accelerated by the field to an energy $W=eE/\alpha$. But the creation of one pair of ions still consumes at least the energy W_{min} (*Stoletov's constant*), which is several times the ionization potential. Electrons have to devote much energy to the excitation of atoms: In air, W_{min} = 66 eV/pair of ions for $(E/p)_m$ = 274 V/(Pa×m). A more accurate value for α is found with the equation

$$\alpha = 1400 \cdot \delta \cdot \left(\frac{E}{3.1 \times 10^8 \delta} - 1\right)^2 \text{ m}^{-1} \qquad (3.12)$$

where δ is the ratio of air density to the normal density ($p = 10^5$ Pa, $T = 25$ °C) [RAI87].

Few accurate measurements of Townsend coefficients have been published, often over only limited electric field ranges. It is therefore rarely practical to use measured Townsend coefficients. Software for calculating such coefficients exists, see for example [CEN99].

γ depends on E/p in an irregular manner and is very sensitive to the state of the cathode surface. The data on γ are incomplete and often contradictory. According to Raizer [RAI87] as a rule, one assumes that $\gamma \sim 10^{-1}$-10^{-2}, while $\gamma \sim 0$-0.3 is possible. Osmokrovic [OSM94] on the other hand states that γ varies from 10^{-4} to 10^{-8}. Electronegative gases (SF$_6$, Freon, O$_2$, CO$_2$) reattach the electrons very quickly, so they have low γ. Lux [LUX01] states that for nitrogen, gamma ranges between 10^{-3} and 10^{-2} for E/p of 75-500 V/(Pa×m). Insulating gases like SF$_6$ or Freon have γ of 10^{-4} or even less.

The increase of E_b in small gaps

The electric field at breakdown becomes large at the extremes of high and low pressure and for an extremely short gap. Fig. 28 on page 69 shows an illustration of this effect with our measured data. The yield of electrons per ion incident on the cathode, γ, is then orders of magnitude larger than that usually encountered for breakdowns under more moderate conditions. Ions may not be made by electron impact with gas molecules in the gap, but arise directly from electron impact with the (contaminated) anode surface [KIS59]. This supports the vacuum breakdown mechanism put forward by Raizer [RAI87]. When breakdown occurs at fields insufficient for ejecting an electron from the metal, a spurious electron may be accelerated in the field, knock an ion from the anode, or emit a bremsstrahlung photon. The ion or the photon knock out, in turn, an electron from the cathode, etc. This multiplication proceeds without the residual gas.

According to Crowley [CRO86] breakdown in air is possible from 3×10^6 V/m, no corona and sparking will be observed below 300 V. Since, as a rule, both limits must be exceeded before breakdown, the voltage and field are related by a characteristic length, the *critical size*, which separates voltage controlled breakdown from field controlled breakdown. This size d_{cr} is approximately

$$d_{cr} = \frac{V_b}{E_b} = \frac{300}{3 \times 10^6} \text{ mm} = 0.1 \text{ mm} \tag{3.13}$$

V_b: breakdown voltage, E_b: breakdown field strength in air.

Below this size, the fields may be large, but breakdown will not occur until the voltage exceeds approximately 300 V. Above this size, the voltage may be large, but the field must exceed 3 MV/m for breakdown.

Vacuum discharge

If $pd < 10^{-3}$ Pa×m, an electron crosses the gap practically without collisions. Latham [LAT81] found that breakdown is independent of p in the range 10^{-6}-10^{-3} Pa for mm gaps. Still, the development of the vacuum discharge occurs in a vapour medium. The vapour is then composed of atoms of one or both of the electrodes. In the final stage of the growth of the discharge, i.e., the arc phase, it is known that the current is sustained in cathode vapour. For breakdown in undegassed electrodes in moderate vacuum it is believed that the growth of current occurs in gas desorbed from electrode surfaces. For small electrode separations, the initial production of electrode vapour necessary for breakdown initiation depends upon interactions of the prebreakdown field-emission current with the electrode surface. Such current heats the surface of the cathode by the Joule effect and the anode by electron impact. Davies [DAV73] associates breakdown with the thermal instability of a region on the anode. It is assumed that breakdown occurs when the surface temperature of the cathode or the anode reaches the melting point of the electrode material. Muller [MUL37] mentions a critical emission current density of 10 A/m^2 (=1 pA/μm^2) where this happens. Beckey [BEC71] gives a theoretical E_b value for a "perfect" vacuum gap of about 6.5×10^9 V/m, whilst the threshold for emission of electrons is about 3×10^9 V/m.

Observations of the luminosity between electrodes during the growth of current from the field-emission value to the arc value show that for very small overvoltages the luminosity appears first at the anode and subsequently bridges the space between the electrodes. However, studies with improved time resolution indicate that the luminosity appears first at the cathode. The possibility that the luminosity is due to excitation of desorbed gas cannot be precluded.

Field Emission of Electrons

For the initiation of a discharge, seed electrons are necessary. At high voltage, natural electrons can be sufficient to start an avalanche in a gas. At high field strength, electrons are emitted from the cathode, thus delivering free electrons as well as heating cathode and anode. Electron emission can also have other, potentially combined, causes: heat, photons, particle impact. Field emission is a major contribution to the pre-breakdown current in our case. It is here the most important factor in breakdown initiation. A method to calculate this emission has been conceived by Fowler and Nordheim [FOW28]. The original Fowler-Nordheim equation was erroneous and has been corrected and developed further by the inventors themselves and other authors. The version we eventually found from our investigations is the following:

$$I = A\cdot\left(\beta\cdot\frac{V}{d}\right)^2 \cdot \frac{q^3\sqrt{(\mu/\phi)}}{2\pi h(\mu+\phi)} \cdot \exp\left(\frac{-8\pi\sqrt{2m\phi^3}}{3hq(\beta V/d)}\right) \qquad (3.14)$$

A: emission area, β: field amplification factor, V: applied voltage between two plates, d: distance between these plates. q: electron charge, m: electron mass, h: Planck constant, Φ: work function of the bulk, μ: chemical potential.

This equation is simplified. It does not take account of the temperature or space charge, nor the so-called *image effect*. It is also difficult to know the exact values for the work function and chemical potential. The use is therefore limited to an order of magnitude estimate of electron emission for a certain geometry, unless the parameters can be found beforehand experimentally. Once an emission curve has been measured, the data can advantageously be graphed in the *Fowler-Nordheim plot*. This plot can serve as a check whether the measured current is in fact emission current, and from the plot β and A are found. To do so, the ordinary Fowler-Nordheim equation (3.14) is rewritten as

$$I/V^2 = A\cdot\beta^2\cdot a\cdot\exp\left(-\frac{b}{\beta\cdot V}\right)$$
$$\Leftrightarrow \ln\left(I/V^2\right) = \ln A + \ln\left(\beta^2\cdot a\right) - 1/V\cdot 1/\beta\cdot b \qquad (3.15)$$

a and b are constants found or estimated from materials and coarse geometry. When s is the slope and i_y the interception point with the y-axis, both found from the linear range of the corresponding F-N plot, β and A are then calculated using

$$\beta = -\frac{1}{s}\frac{8\pi d}{3hq}\sqrt{2m\phi^3} \qquad (3.16)$$

$$A = \exp(i_y) \bigg/ \left(\frac{\beta^2}{d^2} \cdot \frac{q^3\sqrt{(\mu/\phi)}}{2\pi h(\mu + \phi)} \right) \qquad (3.17)$$

Towards high voltages, the curve tends to stall because of accumulating space charge. For a more in depth coverage of the subject of electron emission, the book by Sedlaceck [SED96] is recommended.

The electronic work function Φ

Energy (or work) required to withdraw an electron completely from a metal surface. This energy is a measure of how tightly a particular metal holds its electrons, i.e. of how much lower the electron's energy is when present within the metal than when completely free. The work function is the solid-state analogue of the ionization potential in atoms or molecules. [GOM61]

Table 6 Work functions for a selection of relevant microsystems materials.	Al multicrystalline	4.28	Pt	5.65
	Au multicrystalline	5.10	Si (n-doped)	4.85
	Cr	4.50	Si (p-doped)	4.91
	Cu multicrystalline	4.65	Ti	4.33
	Mo multicrystalline	4.60	W multicrystalline	4.55
	Ni	5.15	C	5.00

The chemical potential μ

The chemical potential of a thermodynamic system is the change in the energy of the system when an additional constituent particle is introduced, with the entropy and volume held fixed. If a system contains more than one species of particle, there is a separate chemical potential associated with each species, defined as the change in energy when the number of particles *of that species* is increased by one.

Since the chemical potential is a thermodynamic quantity, it is defined independently of the microscopic behaviour of the system, i.e. the properties of the constituent particles. However, some systems contain important variables that are equivalent to the chemical potential. In Fermi gases and Fermi liquids, the chemical potential at zero temperature is equivalent to the Fermi energy. In electronic systems, the chemical potential is equivalent to the negative of the electrical potential. [WIK03]

3 Ionization of gases

The field amplification factor β

In order to calculate the maximum field strength near electrode surfaces, which is important for electron emission calculation, the *field amplification factor* β is used to account for the micro geometry of the surfaces:

$$E = \beta \cdot \frac{V}{d} \qquad (3.18)$$

β depends on the electrode geometry (including surface roughness, small protrusions and particles), but also on the chemical state of the surface, i.e. gases and other substances that are adsorbed to or enclosed near the bulk surface. The amplification by surface protrusions usually lies between 100 and 1000. For extremely sharp and long tips (whiskers, nanotubes) it may exceed 10000. Williams and Williams (1972): for polished electrodes, the value of β is in the range 80-130. One way to calculate β is by using the simple equation

$$\beta = 2 + \frac{h}{r} \qquad (3.19)$$

where r is the tip radius and h the tip height, $h/r > \approx 5$. Taking the example of a nanotube with a radius of e.g. 5 nm, β would be about 1000 if the geometry is approximated by a spherical tip on a cylindrical shaft that is $h = 5$ μm high. What becomes important in our special case is the reduction of the β-factor in smaller gaps. To account for the influence of d, an extension leads to

$$\beta = \beta_\infty \left(1 - \frac{h}{d}\right) \qquad (3.20)$$

where β_∞ is the value of β when $d \gg h$ [MIL67]. For large distances, this extension is only a minor correction. But for the low distances of our interest, eq. (3.20) shows that if d is less than $10h$, the amplification effect is already reduced by about 10%. For nanotubes this means $d \geq 50$ μm.

To estimate the real field strength in different structures and to verify the above mentioned calculations for the β-factor for small gaps, computer simulations were carried out. The finite element simulation program Ansys® (Ansys Inc.) was used to simulate the field near the tip of a nanotube for a distance between the electrodes d of 1-500 μm. The tube was assumed to be 5 μm high (h), to have a radius of 5 nm and a spherical tip. The results for the calculated (eq. (3.20)) and simulated maximum field strengths are depicted in Fig. 10.

3.2 Electrical discharges in gases and in vacuum

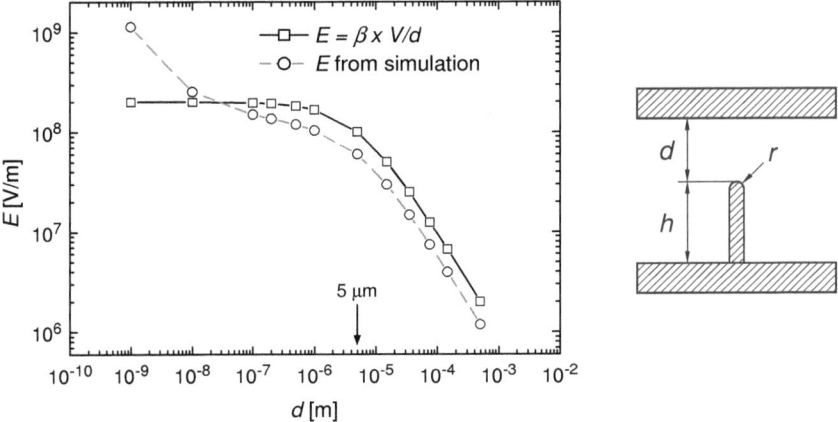

Fig. 10 Simulation of the electrical field strength E near a cylindrical protrusion (e.g. nanotube) compared to the field strength calculated with (3.20). $V = 1$ V, $h = 5$ μm.

Compared to values calculated with the fore mentioned equations, the simulated maximum field strength E, and therefore β, was about 1.7 times smaller for tip distances of 500 μm down to about 5 μm. In this range E is proportional to $1/d$, as expected for parallel plates. In smaller gaps the field amplification effect of the tip decreases while the field strength tends to infinity. Below 5 μm (here: $h = d$), eq. (3.20) is no longer valid. The difference between eq. (3.20) and the simulation above 5 μm is easily explained by the method used by Miller [MIL67] in deriving his equation: As a simplification he used a sphere instead of a cylinder with a spherical tip, which results in a higher field strength.

Additional simulations were done for a rough estimate of the effect of the gross shape of the electrode on which the nanotube or a similar protrusion resides. Two extreme cases were considered: A nanotube on the spherical top of a larger cylinder with a radius of 15 nm and a height of 10 μm, and a nanotube on just the spherical top (which resembles a sphere). The electrode distances were 10 μm and 40 μm. The "large" cylinder without protrusion has an amplification factor of about 300. At an electrode distance of 40 μm this causes an additional amplification of the field at the nanotube tip of only 3.7 times. For just the tip as a base for the tube, the additional factor is 5.4. The simulation at 10 μm gave amplifications of 1.6 and 1.8 times respectively. Compared to the amplification by the nanotube alone, the effect is therefore found to be quite weak. For conical or edge shaped electrodes it will be even lower. These results show how the gross electrode shape becomes less relevant at smaller electrode distances when significant micro protrusions are present. According to Latham [LAT81] it has also been shown experimentally that β (apparently) increases with d. This is called the total voltage effect.

According to the literature, β can easily range from 10-1000. For polished electrodes, a value of β is in the range 80-130 can be expected. For the emission area, a range from 10^{-16} to 10^{-12} m^2 (100 nm^2

3 Ionization of gases

$< A < 10^6$ nm^2) has been found [GOM61, LAT81, REE97]. By Bonard [BON98] field amplification factors of up to 50000 were calculated for single nanotubes at 1 mm interelectrode distance using a method of Dyke and Dolan [DYK56].

3.2.2 Types of DC discharges

Before breakdown, the current over a gap between electrodes will be very low and depend on the processes illustrated in Fig. 8 on page 24. Once at least the breakdown voltage is applied, breakdown occurs and leads to a discharge. The possible forms of discharges and their determining parameters are described in this section. Fig. 11 shows a general characteristic curve for discharges. Following this curve, increasing the voltage applied to the circuit, thus from low to higher currents, we will briefly introduce the discharge types.

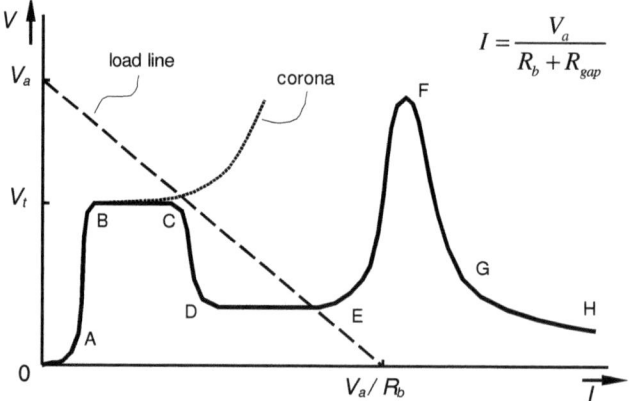

Fig. 11 Current-voltage characteristics of discharges between electrodes for a wide range of currents, and the load line: A-B region of non-self-sustaining discharge, B-C Townsend dark discharge, D-E normal glow discharge, E-F abnormal glow discharge, F-G transition to arcing, G-H arc. [RAI87]

The dotted line shows how the current develops in a highly inhomogeneous field, where breakdown leads to a corona discharge instead of a simple glow discharge as in a homogeneous field. [NAS71]

Table 7 gives an overview of the types of discharges covered here, and their most important characteristics.

Table 7 Overview of types of discharges and electrode current ranges below and at breakdown.

Type of discharge	Characteristics/Conditions/Comments	Current range	Degree of ionization
Cosmic rays (not a discharge)	Low rate, important before Townsend discharge ~10^9 m^{-3}s^{-1} [VEL01], ~10^7 m^{-3}s^{-1} [RAI87]	<< pA	<<
Townsend dark discharge	None or very little light emission Cathode fall of several 100 V [WAS72b]	10^{-10}-10^{-5} A (100 pA - 10 µA)	<<
Corona discharge	Positive and negative corona May be intermittent [RAI87, WAS72b]	About 1 µA Pulses: 100-350 µA	<< in gap, like glow near tip
Glow discharge	May be stable or pulsed Sharply nonuniform distribution of the potential difference across the gap Cathode fall of several 100 V [WAS72b]	~10^{-6}-1 A (1 µA - 1 A)	10^{-8} [RAI87]
Low temperature non thermal plasma	Large temperature difference between electrons and ions	j~mA/cm^2	10^{-3}-10^{-4}
Arc discharge	When I ~ 1 A, the glow discharge cascades down to an arc. [RAI87]	$\geq$ 1 A	>>

The circuit to which the voltage is applied is basically the one depicted in Fig. 12. The indicated vacuum system allows to control the gas type and pressure in the gap. The ballast resistor stands for the more complicated loading impedance of the whole circuit.

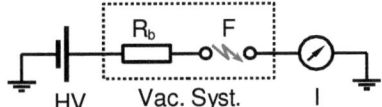

Fig. 12 Simplified circuit for discharge experiments. HV: high voltage source, F: discharge gap, R_b: ballast resistor, I: current meter.

When a voltage is applied to the circuit, the gap voltage V_g can be calculated with

$$V_g = V_a - R_b I \qquad (3.21)$$

where V_a: applied voltage (also called the *electromotive force*), R_b: ballast resistance (more precisely: loading impedance), I: circuit current.

Up to the (interelectrode) voltage V_t, almost no current is measured ($I < 10^{-10}$ A). The current is proportional to V until it reaches saturation: all randomly appearing electrons reach the anode. At $V_a = V_t$ an abrupt transition to a self-sustained current occurs that leads to a drop of the interelectrode voltage. Because of this transition, V_t is also called *transition voltage*. Depending on the electrode geometry and circuit resistance, V_t can coincide with the breakdown voltage V_b. $V_b = V_t$ in homogeneous fields. The discharge type is largely determined by the discharge current scale (the value of current dictates the degree of gas ionization). Therefore the ohmic resistance that limits the current imposes the type of discharge that is produced after breakdown. [RAI87]

In order to measure a *V-I* curve, either V_a or R must be varied and I measured. The load line in Fig. 11 shows the *V-I* relationship in the circuit when the voltage is fixed. At a given voltage V_a, the current I will take the value determined by the sum of the ballast and gap resistance, R_b+R_g. At breakdown, the current increases and the gap voltage decreases along this load line. This "breakdown"-line stops at a point defined by the gap resistance, i.e. where the load line crosses the *V-I* curve. Depending on the *V-I* characteristics, the applied voltage and R_b the load line may cross the curve at several points. Then the situation is not stable and the discharge type can change randomly. Looking at the curve it must be expected that discharges are much more stable using higher ballast resistances, because the load line then becomes steeper.

Townsend Discharge

The Townsend dark discharge is self sustained. This discharge is made self-sustained by applying to the electrodes the voltage equal to the *ignition potential* V_t. This voltage ensures the stationary reproduction of electrons ejected from the cathode and pulled to the anode. Charges are multiplied in avalanches. Practically the entire space is charged positively, but weakly. The current is usually between 10^{-10} and 10^{-5} A (100 pA - 10 µA), depending on the circuit and the electrode surface. The voltage across the electrodes begins to decrease after a certain current is reached, that is point C in Fig. 11, the transition to glow [RAI87]. For inhomogeneous fields, the discharge is called Townsend-like and has a smooth transition to corona if the tip is negative [HEL97].

The Townsend discharge is dark because at this stage excitement of atoms by electron impact is not important and ionization is so small that the gas emits no appreciable light. The number of charges in the gap is to the greatest part determined by impact ionization. The number is not great enough to cause a space charge that would noticeably change the potential distribution in the field. [WAS72a]

The reason for the horizontal line B-C Fig. 11 is similar as for the horizontal voltage line at glow discharge. The current range for this line depends on the emission area, i.e. the area where field strength is similar on the electrode surface. A single tip has a low area, which means a short current range for a Townsend discharge. Then it is difficult to find the correct ballast resistance and voltage for operation in the Townsend regime. Since the electrode state is not constant, operation near either of the edges above or below Townsend results in an unstable current.

Glow Discharge

The main distinction of the glow discharge from the dark discharge lies in a sharply nonuniform distribution of the potential difference applied across the gap to the electrodes. [RAI87] The glow discharge can be divided into several sub-categories. The most important ones that also have their regions in Fig. 11 are subnormal, normal, and abnormal. The obstructed and the hollow cathode discharges are special cases of glow discharges that are also mentioned here. Glow discharge will be covered in detail in chapter 4 .

Subnormal Glow Discharge

This is a transition region between the glow and dark discharge regions that corresponds to currents so weak that the size of the "quasinormal" cathode spot is found to be comparable to the *cathode layer* thickness. The loss of charges in the lateral direction is harmful for multiplication, so that the voltage across the layer required for self-sustainment of the discharge is found to be higher than for the normal regime. [RAI87]

Normal Glow Discharge

In a glow discharge the cold cathode is emitting electrons due to secondary emission mostly due to positive ion bombardment. A distinctive feature of this discharge is a layer of large positive space charge at the cathode, with a strong field at the surface and considerable potential drop of 100-400 V (or more). This drop is known as *cathode fall*, and the thickness of the cathode fall layer is inversely proportional to the density (pressure) of the gas. If the interelectrode separation is sufficiently large, an electrically neutral plasma region with fairly weak field is formed between the cathode layer and the anode. Its relatively homogeneous middle part is called the *positive column*. It is separated from the anode by the *anode layer*. The current range in the normal mode is the greater, the higher the pressure. The current range is also greater in a longer the tube, when the discharge takes place in a tube [RAI87].

Abnormal Glow Discharge

After the entire cathode has been covered with discharge, any further increase of current inevitably increases the density at the cathode in comparison with the normal value. This region of increase current density is called the *abnormal* glow discharge. In actual situations, a cathode fall of more than several kV and current densities of order 10^5-10^6 A/m^2 (0.1 - 1 µA/µm^2) result in intense heating of the cathode and transition to an arc discharge. [RAI87]

Obstructed discharge

This mode arises at very low pressure in narrow gaps of widths d, such that the product pd is less than the normal layer thickness pd_n. Roughly speaking, these conditions correspond to the left-hand branch of the Paschen curve, where $V > V_{min}$. The interelectrode separation is insufficient for "normal" multiplication, so that voltage has to be raised in comparison with the normal value. If this is not possible the discharge is extinguished. [RAI87]

3 Ionization of gases

Hollow cathode discharge

If the cathode is arranged as two parallel plates (the anode being shifted to the side) and the cathode plates are brought closer and closer, the current increases hundreds and thousands of times after a certain distance is reached. This takes place when two formerly nonoverlapping regions of negative glow merge: the glow becomes considerably more intensive, and the voltage changes slightly. A similar effect can be obtained if the cathode is a hollow cylinder and the anode lies far along the axis. The pressure must be such that the cathode layer thickness is comparable with the cylinder diameter. In a hollow cathode, electron streams converge to the axis and produce intensive ionization and excitation of the gas. Photoemission excited on the cathode by UV radiation produced in this region also plays a role here. [RAI87]

Thermal effects are either absent or not determining for the discharge maintenance. The glow discharge belongs to the right hand side of the Paschen breakdown curve, near the minimum for a pressure range 0.01-1 kPa, burning voltage 100-500 V and current density 0.1-1000 A/m^2. The "normal" glow belongs to a current interval 10^{-4}-10^{-2} A in which the current-voltage characteristic has a plateau.

When the cathode encloses a hollow space, the negative glow is confined inside the cathode for a specific range of operating conditions. Then at a constant current the burning voltage is found to be lower than in the case of plane electrode geometry, whereas at constant burning voltage the current increases several orders of magnitude. It was Paschen who called this effect a "hollow cathode effect". The hollow cathode effect is observed only in a limited range of pd, 10-100 Pa×m, where p is the gas pressure and d is the diameter of the cathode cavity. A necessary condition for the hollow cathode effect to occur is the characteristic dimension d to exceed the mean free path of the primary electrons at least by an order of magnitude. The hollow cathode discharge characteristics are also influenced by other geometrical parameters such as the shape and length of the cathode, whether it is opened or closed, the shape and the distance of the anode and the insulating walls.

Corona Discharge

The corona discharge appears in inhomogeneous fields. Details about this partial discharge are covered in detail in the next chapter.

Arc Discharge

The discharge known as "the arc" has a relatively low cathode potential fall (in the order of the ionization or excitation potential of atoms, that is, about 10 eV). This characteristic distinguishes the arc discharge from the glow discharge, in which the cathode fall is hundreds of volts. The small cathode fall results from cathode emission mechanisms that differ from those in the glow discharge. These mechanisms are capable of supplying a greater electron current from the cathode, nearly equal to the total discharge current. This factor eliminates the need for considerable amplification of the electron current, which is the function fulfilled by the high cathode fall in glow discharges. In an

3.2 Electrical discharges in gases and in vacuum

arc, the cathode emits electrons as a result of *thermionic, field electron, and thermionic field emission*. The arc discharge is characterized by large currents much greater than the typical currents of glow discharge. Arcs usually burn at low voltage not exceeding 20-30 V for short arcs, and in some cases as low as several volts. Arc cathodes receive large amounts of energy from the current and reach high temperatures, either over the entire cathode area or just locally, usually for short time intervals. They are eroded and suffer vaporization. [RAI87]

As on the cathode, the arc may be anchored to the anode in two ways:

a) Diffuse anchoring. In this mode, the current is spread over a relatively large area of the anode, at a density $j \approx 10^6$ A/m^2. Material erosion is negligible in this mode because energy flux densities at the surface are not very high.

b) Anode spots. A spot is formed when the anode is small and the growing current is forced to occupy its edges, "awkward" areas, lateral patches, etc. At a certain current the discharge is destabilized and contracts at the anode surface. Sometimes many spots are formed, arranged in symmetric regular patterns. The spots may move. They are very bright and eject vapour jets.

Arcing is the usual condition of atmospheric pressure plasmas and special circumstances must be developed in order to prevent arcing. Helium is effective in preventing arcing because it has a low first Townsend coefficient (ionization rate), such that Helium discharges naturally have a high impedance, which limits the current flow through the plasma. This helps limit arcing, due to the high impedance of the plasma. Arcing frequently is caused by localized emission of electrons, which reduce the discharge impedance and then increase the local ion current, resulting in a run-away situation. [SEL97]

Streamers and Sparks

An avalanche transforms into a *streamer* when two conditions are met: 1. applied field above the so-called critical field strength, E_{cr}, and 2. sufficient or *critical distance*, d_{cr}, between the electrodes. Streamers tend to appear especially during the initial phase of a breakdown. The critical field in ambient air is roughly 3×10^6 V/m. At this field strength the ionization coefficient equals the attachment coefficient. The critical distance is a fraction of a millimetre. In a streamer, the electric field of the ions that are left behind dominates the applied field [GEO01]. The process of initialization of breakdown from avalanche to streamer is illustrated in Fig. 13.

3 Ionization of gases

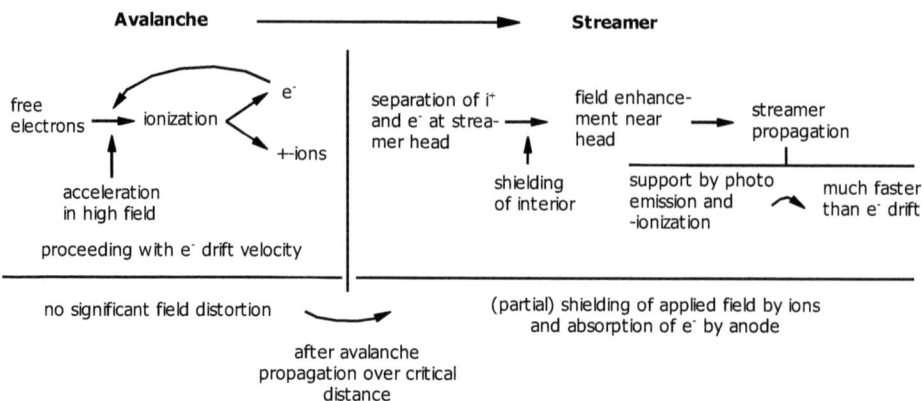

Fig. 13 Illustration of the formation of streamers from an avalanche when the conditions for streamer formation are fulfilled: 1. applied field above the so-called critical field strength, E_{cr}, 2. sufficient or critical distance, d_{cr}, between the electrodes.

As long as the net charge is not sufficient to distort the field appreciably, the centre of the avalanche moves with the electron drift velocity appropriate to the applied field. When the electron avalanche grows to a size such that it is capable of partially shielding itself from the applied field, the propagation and growth of the avalanche change markedly, and the streamer phase follows. Essentially, a streamer is an ionization wave. In front of the wave (the streamer head) the separation of positive and negative charge particles would shield the interior, and cause a sharp enhancement of the electric field over a limited region just outside the streamer head. For fields near that required for breakdown, the ionization coefficient is a strong function of the electric field, so that even a modest field enhancement can result in substantial increases in the ionization rate. If a mechanism, such as photoemission of photoionization, exists that places a few free seed electrons just in front of the streamer head, the effect of an avalanche in the locally enhanced field can cause the streamer to propagate with velocities much higher than the peak electron drift velocity. According to a simulation by Georghiou [GEO01], a streamer will cross a 100 μm gap within a few ns. Therefore a 10 μm gap would be crossed in less than a ns, if the same mechanism appears. Van Veldhuizen [VEL02] mentions a range for streamer velocity from 10^5 to 10^6 m/s, i.e. 0.1 to 1 μm/ps. A typical value for the velocity of electrons is 50 μm/ns [BAT97]. Since most processes in the development of a discharge are related to collisions, the time of evolution is reduced as the gas density increases [RAI87].

A streamer can create a conducting path between the electrodes through which a current will flow. If this current is very low then recombination will reduce the conductivity and the discharge will extinguish. A high enough current will heat the gas, decrease its density and increase its conductivity. The current will grow to a value that is determined by the power circuit. If the power circuit maintains this high current only for a short time, this type of discharge is called *spark breakdown*. Sparks usually have a duration in the microsecond range. Even shorter discharges can be created which in fact stop before they are completed in to arcs. These are sometimes called *transient discharges* and have typical time duration in the range of nanoseconds [VEL02]. In small gaps streamers will always complete and lead to a short circuit, only broken by space charges in the gap.

In discharge gaps, sparks can have a beneficial effect. Since sparks tend to develop from points of high field strength, i.e. tips, such tips are destroyed by sparking. If the energy of the sparks is limited, by a high circuit or plasma resistivity, or the build up of space charges, tips will be rounded and conditioning of the gap is achieved. Sparks that are more violent will also destroy sites of high field strength, but create new ones in the process, or even destroy the electrodes completely, which is especially the case for thin film electrodes.

3.3 Choice of an ionization method

3.3.1 A comparison of ionization methods

Our ionizer is supposed to serve as an ion source. Such sources have the following important general characteristics as presented by Roboz [ROB68]:

- energy spread
- sensitivity
- ionic species produced
- background - from matrix ions, electron current, etc.
- memory - adsorption/desorption to/from surfaces
- mass discrimination - done by filter
- ion current stability, short and long-term

Some of these characteristics have an impact on the possible resolution, some rather on the selectivity of the device. The most important additional characteristics for the purposes of our project are:

- a measurable number of ions must reach the detector
  (lowest reasonably measurable current according to our experience: 10 pA)
- easy to realize the method in a microsystem
- availability of components
- low power consumption
- running at high pressure (preferably atmospheric)
- simple possibility to extract ions to a filter

The following table (Table 8) lists ionization methods with their advantages and disadvantages with respect to our project.

Table 8 Rating of ionization methods from point of view of our research interest and possibilities.

Ionization method	Advantages	Disadvantages	Rating	Reference
Discharge ionization				
Corona discharge	High ion current. Additional info from switching polarity	For stable operation: dry matrix gas. Erosion of electrodes. High ΔE between electrodes necessary: large d. Neg. corona: pos. ions drawn to tip → difficult to extract ions; pos. corona: high voltage, streamers	o	[RAI87] pulsed: [XU01]
DC glow discharge	High ion current. Simple driving electronics and operation	For stable operation: dry matrix gas. Erosion of electrodes	+	[RAI87]
RF glow discharge (capacitively or inductively coupled)	High ion current. Electrodes can be insulated / protected: long term stability	Sophisticated driving electronics necessary	o	[GES00, RAI87, SCP00]
Thermal ionization		Works at high temperature. High power consumption	-	[http06a]
Radioactive ionization[*]	No power supply, low noise level	Radioactive hazard	-	
Particle bombardment	Soft method	For non-volatile compounds needs additional ion source	-	[http01]
Atmospheric pressure ionization (spray methods)	Soft method	For non-volatile soluble components	-	[http01]
Chemical ionization[*]	Produces little fragmentation ("soft" method). (Specific to certain analytes)	Reagent ions necessary: additional ionizer. (Specific to certain analytes). Thermally stable, volatile compounds only	-/o	[http01]
Photo ionization[**]	For IMS: creates short and small ion pulses	Needs strong UV or laser light	-	[RAI87]
Associative ionization, ionization by excited atoms[**]	–	–	-	[RAI87]

3.3 Choice of an ionization method

Ionization method	Advantages	Disadvantages	Rating	Reference
Atmospheric pressure chemical ionization	Soft method	Like ESI, but more fragmentation Additional corona discharge necessary	-	[http01]
Other methods				
Field ionization	Soft method	Very high E and ΔE between anode and cathode necessary: large d Strong emission of secondary electrons adds to total I Highly dependent on surface state and adsorbed layers	-	[BEC71]
Electron (impact) ionization[**]	Well established in mass spectrometers	Separate electron source necessary. For low molecular masses only high fragmentation	o	[http01] [PET00]
Flame ionization		Specific to hydrocarbons	-	[ZIM01]

[*] Used in IMS; [**] Found in discharges

3.3.2 Industrial ion sources

Ion sources generate ion beams with defined energy, current, and energy distribution. Many different ion sources have been developed; industrially important ones are sketched in Fig. 14. All have in common a cold or heated cathode in a high-frequency or stationary electric and/or magnetic field in which a gas discharge is created by electrons that ionize gas molecules. Neutral gas molecules can be introduced into the source, a liquid or a solid can be vaporized, or neutral atoms or molecules can be obtained by sputtering. The ions are removed from the discharge by an extraction electrode connected to a desired acceleration voltage, and the ion beam is formed by electrostatic lenses. [SED96]

The industrially important ion sources illustrated in Fig. 14 are:

- Ion source with heated cathode (a)
- Penning source (b)
- High-frequency ion source (c)
- Ion source with vaporization (d)
- Duoplasmatron (e)
- Ion source with grid extraction (f)
- Electron Cyclotron Resonance (ECR) ion source (g)
- Electron beam ion source (EBIS) with potential distribution along the axis (h)

3 Ionization of gases

Most of these sources are too complex to be realised as a microsystem: some contain permanent magnets and coils, while integration of such needs expertise in itself.

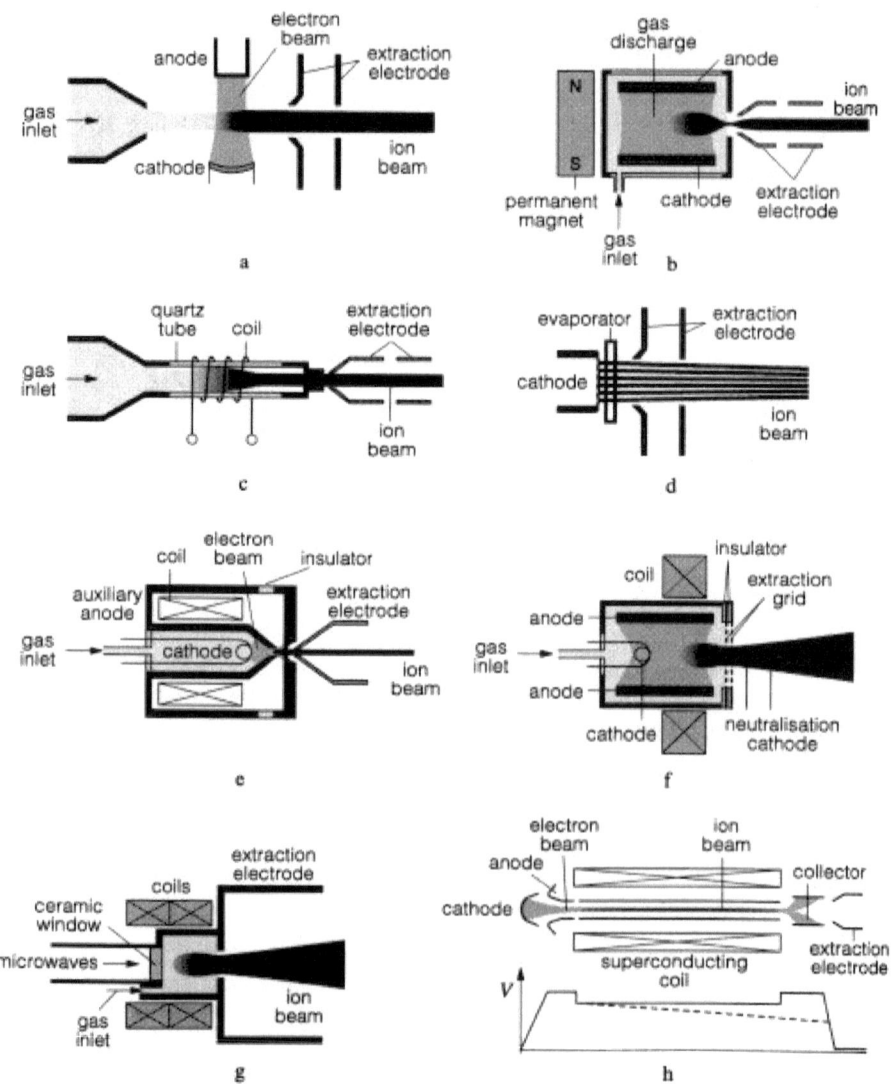

Fig. 14   Examples of industrial ion sources. [SED96]

3.3.3 Choice of a method

The ionization methods that are not readily eliminated by the above comparison are:

- electron (impact) ionization
- discharge ionization: corona discharge, DC glow discharge, RF glow discharge (capacitively or inductively coupled)

For ionization with an *electron beam*, a separate electron source is necessary. Petzold [PET00] reported on a micro ionizer that uses an RF plasma as an electron source. This plasma runs at 100 Pa, probably in Ar. This system seems quite complex, but feasibility is proven.

To ionize directly in a plasma seems somewhat simpler. *Glow discharge* is routinely used as a means of ionization in mass spectrometers, although no application of the method in a micro system is known.

In a microsystem, *inductively coupled discharge* has been realized [HOP00]. Scheffler et al. [SCP00] have fabricated a microstructured planar *RF plasma generator*, while Schoenbach et al. [SCK95, SCK97], Frame et al. [FRA97], and Biborosch [BIB99] worked on *micro hollow cathode discharges*. Yasuoka et al. [YAS01] published their work on pulsed operation of micro-hollow cathode plasmas. Using a hollow cathode structure increases the degree of ionization in the discharge significantly.

An example for *corona discharge* in a true micro system has not been found so far. This is not surprising, since corona discharge needs a highly non-symmetrical electrode configuration that is difficult to realize with a short electrode distance. As found from our own examination of *field ionization* [LON01], a minimum electrode distance must also be maintained for corona discharges. An example for application of a pulsed corona discharge as an ion source for a miniaturized ion mobility spectrometer has been presented by Xu et al. [XU01]. When using sharp tips, there is the danger of tip erosion, which can only partly and for a short term be compensated for by applying a large number of tips. Unlike field emission tips, tips in a corona discharge are in direct contact with a glow, consisting partly of high energy particles, continuously hitting the surface.

Applying glow discharge in DC mode also bears the risk of damaging the cathode by sputtering, as is demonstrated in the work of Eijkel et al. [EIJ00c]. In spite of sputtering, lifetime can be long if the current is kept low and/or if the discharge is pulsed. Working at high pressures is an advantage here, because collisions reduce the energy of particles in a discharge.

The much simpler fabrication of a glow device is one advantage for glow compared to corona. Corona is made more interesting by the possible application of nanotubes or similar nanostructures. A disadvantage of a micro corona device is the lifetime that is expected to be rather short because any sharp tip is likely to burn quickly in the O_2 containing atmosphere in which we may want to work.

In an RF discharge, sputtering can be almost completely avoided, and even a protective insulating layer may be applied [SCP00]. Such a layer makes the device much less sensitive to the influence of measurand gases or other environmental impairments. From such a device the best long term performance can be expected. The disadvantage of RF plasma is the more complicated driving electronics. Employment of standard RF components as applied for example in mobile phones—if feasible—renders RF discharges attractive.

Inductively coupled discharges become interesting to explore if a suitable micro coil is at hand. The electronic part is comparable in complexity to capacitive coupling, the advantages are the same.

The decision

RF discharges apparently require a great degree of additional expertise for the discharge theory as well as for the peripheral electronic devices. Seen the resources and the experience available for the project, discharge ionization appeared to be most attractive. Working on DC discharges is the straight-forward approach, because the equipment for experiments on hand was sufficient or easily obtainable, and non-destructive glow in Ar and N_2 with a microstructure, as well as miniaturised corona with a grid in air, have already been achieved in preliminary tests. DC micro glow and corona discharges being used for gas analysis are little known in literature, making them scientifically worthwhile to explore.

4 The DC glow discharge

We decided to use a DC glow discharge to ionize the gas to be analysed. Therefore the theory of glow discharges is the basis of our research work regarding the ionizer. This chapter forms a collection of the basics of DC glow discharge, with special consideration of discharges in small gaps.

4.1 Characteristics of DC glow discharges

The glow discharge is a self-sustaining discharge with a cold cathode emitting electrons due to secondary emission, mostly due to positive ion bombardment. A layer of large positive space charge is produced at the cathode, with a strong field near the surface and a considerable potential drop of 100 - 400 V or more, known as the cathode fall. If the interelectrode separation is sufficiently large, an electrically neutral plasma region with fairly weak field is formed between the cathode layer and the anode. Its relatively homogeneous middle part is called the positive column. It is separated from the anode by the anode layer. Various aspects of the dark and bright luminous regions of which a glow discharge consists under ideal circumstances (homogeneous field, sufficient gap length) are shown in Fig. 15 on the next page.

The distances from the cathode to characteristic points are dictated by the number of electron free paths λ, depending on the pressure, $\lambda \propto p^{-1}$. At increasing pressure, all the layers become thinner and shift closer to the cathode. If the electrodes are moved closer to each other at constant pressure, the positive column is shortened. As the electrodes come still closer, the column disappears, then the Faraday space and finally the negative glow vanishes. The cathode layer is vital for the glow discharge. If the distance is insufficient for the formation of the cathode layer, the glow discharge cannot be ignited. The higher the pressure, the wider the current range in which the normal mode is realized. When the discharge takes place in a tube, then the current range is increased with the length of the tube.

Different factors are related to the homogeneity of the plasma. Plasma instability is quite often apparent under visual observation. First of all, the differences in luminosity are caused by unequal electron density. The factors that cause inhomogeneity are related to the processes that control the density of electrons, their production, removal and spatial transfer. The symbolic equation of kinetics of electrons is

$$\frac{dn_e}{dt} = Z_+ - Z_- \tag{4.1}$$

where dn_e/dt is the number of electrons rate and Z_+ and Z_- are the production and removal rates, respectively. The rates depend not only on n_e but also on other parameters like the electron temperature, field, and ion density. The discharge is stabilised by the external resistance, which limits the

4 The DC glow discharge

current. Gas heating is a destabilising factor, as it results in a drop in density, leading to the so-called *thermal instability*. [RAI87]

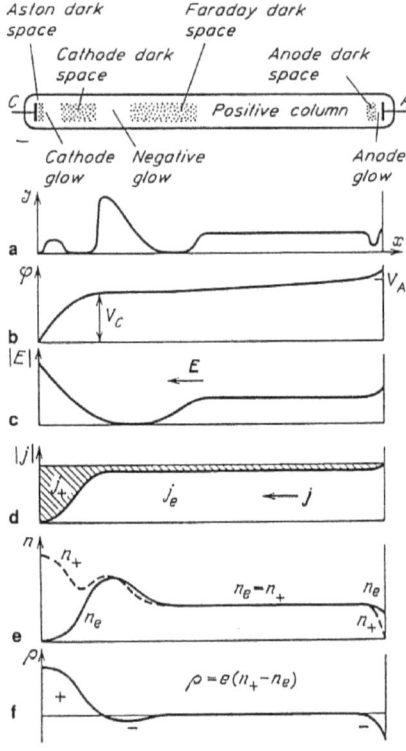

Fig. 15 Qualitative structure of a glow discharge in a tube and the distribution as a function of the distance x from the cathode of: (a) glow intensity J, (b) potential φ, (c) longitudinal field E, (d) electron and ion current densities j_e and j_+, (e) charge densities n_e and n_+, and (f) space charge $\rho = e(n_+ - n_e)$. [RAI87]

4.1.1 The cathode layer and the Debye length

Under ideal circumstances, the cathode layer is an autonomous system consisting of the negative surface charge on the cathode and a boundary layer of positive space charge (*sheath*), extending over a narrow region of space across which the potential changes rapidly. Poisson's equation relates a potential structure to space charge, ρ:

$$\frac{d^2\Phi}{dx^2} = -\frac{\rho}{\varepsilon_0}$$

(4.2)

4.1 Characteristics of DC glow discharges

The scale length, *l*, of the sheath region can be estimated by

$$\frac{\Phi}{l^2} \approx -\frac{n_i e}{\varepsilon_0} \tag{4.3}$$

where *e* is the charge of an electron. With $\Phi \sim kT_e/e$, and the plasma density $n_i \sim n_e$, we find what is known as the *Debye length* λ_D

$$\lambda_D = \sqrt{\frac{\varepsilon_0 kT_e}{n_e e^2}} \tag{4.4}$$

T_e is the electron temperature, *k* is Boltzmann's constant, and n_e is the electron density in the plasma [BRA00]. The Debye length is one of the characteristic properties of plasma. It is defined as the maximum dimension of the space charge region where quasineutrality can be disturbed. A normal glow can only exist if λ_D is smaller than the smallest linear dimension *d* of the region containing the plasma. With the standard values of the constants in the SI system, this simplifies to

$$\lambda_D = 69.01 \sqrt{\frac{T_e}{n_e}} \tag{4.5}$$

The approximate energy range of low pressure glow discharges is from 1.6 eV to 16 eV [TSU01]. Converted to temperature using the relation 1 eV = kT/e, this corresponds to a temperature from ~2×10^4 K to 2×10^5 K. In the high pressure regime we are working in, the electron temperature will be rather low, we therefore assume $T_e = 1 \times 10^4$ K. According to Sedlaceck [SED96], the electron density in a laboratory plasma can vary between 10^{12} m^{-3} in a low pressure plasma and 10^{24} m^{-3} in high pressure arc discharges while the Debye length varies between 10^{-2} m and 10^{-8} m. With $n_e = 10^{18}$ m^{-3}, at the centre of Sedlacecks values, we find a Debye length of about 7 µm. The effect of variations from the assumed values within reasonable limits are shown in Fig. 16. Within a reasonable range for our microplasma, T_e has little effect on λ_D, while the effect of n_e is comparatively large. We find, that in a microgap ≤ 10 µm a glow can exist, if the electron density is greater than about 5×10^{17} m^{-3} if $T_e = 10^4$ K. In a gap that is slightly smaller than λ_D, the glow will become abnormal, i.e. have an increased current density, while in much smaller gaps only arcs will develop.

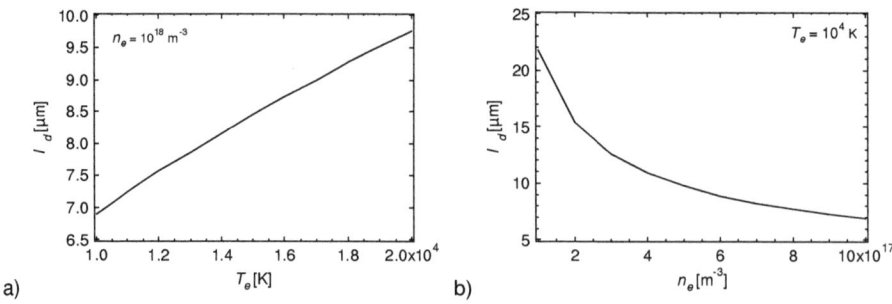

Fig. 16 Comparison of effects of parameter variations on the Debye length λ_D.
a) λ_D over electron temperature T_e at $n_e = 10^{18}$ m^{-3}, b) λ_D over electron density at $T_e = 10^4$ K.

The Debye length depends on the combination of cathode material and gas, because this combination influences the electron density of the plasma at a given voltage. For the normal cathode fall and layer thickness values are found that are close to V_{min} and pd_{min} for the breakdown of a plane discharge gap in the same gas and of the same cathode material. For examples, see Table 9 - Table 11. [RAI87]

Table 9 Normal cathode fall V_n. [RAI87]

gas cathode	air	Ar	He	H_2	Hg	Ne	N_2	O_2	CO	CO_2
Al	229	100	140	170	245	120	180	311	–	–
Ag	280	130	162	216	318	150	233	–	–	–
Au	285	130	165	247	–	158	233	–	–	–
Bi	272	136	137	140	–	–	210	–	–	–
C	–	–	–	240	475	–	–	–	526	–
Cu	370	130	177	214	447	220	208	–	484	460
Fe	269	165	150	250	298	150	215	290	–	–
Hg	–	–	142	–	340	–	226	–	–	–
K	180	64	59	94	–	68	170	–	484	460
Mg	224	119	125	153	–	94	188	310	–	–
Na	200	–	80	185	–	75	178	–	–	–
Ni	226	131	158	211	275	140	197	–	–	–
Pb	207	124	177	223	–	172	210	–	–	–
Pt	277	131	165	276	340	152	216	364	490	475
W	–	–	–	–	305	125	–	–	–	–
Zn	277	119	143	184	–	–	216	354	480	410
glass[a]	310	–	–	260	–	–	–	–	–	–

4.1 Characteristics of DC glow discharges

gas cathode	air	Ar	H$_2$	He	Hg	N$_2$	Ne	O$_2$
Al	0.25	0.29	0.72	1.32	0.33	0.31	0.64	0.24
C	-	-	0.90	-	0.69	-	-	-
Cu	0.23	-	0.80	-	0.60	-	-	-
Fe	0.52	0.33	0.90	1.30	0.34	0.42	0.72	0.31
Mg	-	-	0.61	1.45	-	0.35	-	0.25
Hg	-	-	0.90	-	-	-	-	-
Ni	-	-	0.90	-	-	-	-	-
Pb	-	-	0.84	-	-	-	-	-
Pt	-	-	1.00	-	-	-	-	-
Zn	-	-	0.80	-	-	-	-	-
glass [a]	0.30	-	0.80	-	-	-	-	-

Table 10 Normal cathode layer thickness pd_n (Units: cmTorr). [RAI87]

gas cathode	air	Ar	H$_2$	He	Hg	N$_2$	O$_2$	Ne
Al	330	-	90	-	4	-	-	-
Au	570	-	110	-	-	-	-	-
Cu	240	-	64	-	15	-	-	-
Fe, Ni	-	160	72	2.2	8	400	-	6
Mg	-	20	-	3	-	-	-	5
Pt	-	150	90	5	-	380	550	18
glass [a]	40	-	80	-	-	-	-	-

Table 11 Normal current density j_n/p^2 (Units: μA/cmTorr^2). [RAI87]

Arkhipenko [ARK01] states that the stability of an atmospheric pressure glow discharge with normal current density is problematic. The small spatial sizes of the near electrode layers and the high gradients of temperatures, and high concentrations of charged particles result in essential deviations of the plasma condition of glow discharge from the equilibrium state. Various instabilities develop and, as the result, a glow-to-arc transition takes place. Rather small (some tens of μm) sizes of the near electrode regions of the discharge create considerable experimental difficulties in a study of these regions and in the theoretical modes of cathode processes as well. That is why there are no references to experimental results of the cathode fall region parameters in high pressure gases.

4.1.2 Anode layer

The anode fall is defined as the voltage between the anode and the extrapolated value of the linear potential gradient of the positive column to the anode. This voltage is only a few volts, depends on the shape of the anode and can be influenced by the shape and the distance of a neighbouring grid. The crucial function of the anode layer is the fabrication of ions streaming to the cathode. The ions are generated by electron impact. The greater part of the electron acceleration occurs within about four main free path lengths. The voltage of the anode fall increases with increasing current, and decreases with increasing pressure. [WAS72b]

4 The DC glow discharge

4.2 Corona discharge

The corona discharge is a *partial* breakdown. Field strength, i.e. electron acceleration, is only high enough for avalanche in a limited region between the electrodes. Corona discharges occur only if the field is sharply nonuniform. The field near one or both electrodes must be much stronger than in the rest of the gap. This situation typically arises when the characteristic size of the electrodes is much smaller than the interelectrode distance. Otherwise the increase of voltage between the electrodes produces a spark between them, and not a corona discharge [RAI87].

The use of corona discharges to ionize gases in ion mobility spectrometers was studied in recent years by Tabrizchi and co-workers [KHA01, KHA03, TAB00, TAB99].

4.2.1 Development of a corona discharge

In order to regard inhomogeneous fields the equations for breakdown in homogeneous fields given in §3.2.1 must be extended. This has for example been done by Liu [LIU95] and by Raizer [RAI87]. It must be noted that in inhomogeneous fields at higher pressures *space charges* become significant earlier than in the homogeneous case.

The current density at which the field and discharge structure are considerably modified and which manifests the beginning of dark-to-glow transition of discharge in an inhomogeneous field, is given within an order of magnitude by the formula presented by Raizer [RAI87]

$$\frac{j_L}{p^2} \approx \frac{(\mu_+ \cdot p)(E_t/p)^2}{8\pi(pL)} = \frac{(\mu_+ \cdot p)V_t^2}{8\pi(pL)^3} \qquad (4.6)$$

j: current density, μ: mobility, p: pressure, E_t: transition field strength, L: characteristic length, V_t transition voltage.

When a positive voltage is applied, e.g., to a wire in air, the first corona phenomenon observed is the onset of a streamer from the wire, followed rapidly by the formation of the *"Hermstein glow"*, an apparently continuous glow corona uniformly covering the wire, but not extending as far into the gap as onset streamers [MOR97b]. The corona current is limited by the space charge of the charge carriers in the outer region.

The development steps of a positive corona are:
1. An asymmetric electrode configuration is made
2. A high voltage is applied
3. Some free electric charge is present
4. An avalanche builds up and leaves a space charge area behind
   Onset streamers always precede the formation of the glow!

5. Photons from the avalanche create new charge carriers outside the space charge area
6. New avalanches develop closer to the cathode

For a corona to develop, the field must be inhomogeneous, so its strength is high enough to cause ionization near the tip but not near the counter electrode. A negative corona has the same ignition criteria as a Townsend discharge. In a positive corona, electrons are reproduced by photoprocesses. The ignition of corona under laboratory conditions manifests itself not only by a luminous layer around the electrode (which may not be noticed at all) but also by a jump in the discharge current to about 10^{-6} A. The mechanism of multiplication of electrons is essentially dependent on the polarity of the electrode surrounded by the corona. If this electrode is the cathode, then avalanche multiplication takes place. The secondary process is the emission from the cathode and, possibly, photoionization in the bulk of the gas.

If the wire (or tip) is the anode, the remote large cathode does not participate in multiplication, on account of the weak field in its vicinity. The reproduction of electrons is ensured by secondary photoprocesses in the gas around the tip. In contrast to the homogeneous glow of a negative corona, a positive corona displays luminous filaments running away from the tip. Charge carriers are produced only in the direct vicinity of the corona-carrying electrode surrounded by a strong field. In the remaining part of the gap (outer region), the current is carried by charges that are pulled out by the weak field present there. The carriers are positive ions in positive coronas and negative ions in negative coronas (or electrons if the gas has no electronegative components). [RAI87]

The onset voltage of a positive corona V_{C+} is in general slightly higher than V_{C-}. This difference is gas dependent. For quantitative calculations of the onset voltage of corona, *Peek's law (1929)*, formulated from empirical observations, can be used. For coaxial cylinders at normal density of air:

$$E_c = (31.53 + 0.963/r^{1/2}) \times 10^5 \text{ V/m} \tag{4.7}$$

where E_c is the corona onset field and r is the inner conductor radius in metres [PEP97].

A multitude of processes occurs in the gap during a discharge and shortly after it has stopped:

1. Ionization (α)
2. Attachment (η)
3. Recombination (β)
4. Vibrational relaxation
5. Metastable quenching
6. Radical reactions
7. Electron drift and diffusion
8. Ion drift (positive and negative)
9. Formation of metastable oxygen molecules, detachment, collisional quenching

4 The DC glow discharge

Remarks:

2.: The conductivity of a plasma is strongly influenced by the capturing of low energy electrons. Therefore the field strength for sustaining a certain current is much higher in an electronegative gas. A small amount of water can have a big impact.

3.: The recombination time in the atmospheric corona discharge is estimated to be in the order of 1 µs. The time interval required for a corona pulse in order not to notice its predecessor is found in experiments to be in the order of 1 ms. [MOR97a, RAI87]

4.2.2 Intermittent corona discharge

The *inception time lag*, i.e. the delay between voltage application and start of a discharge, varies in practice from about 1 ns to many µs. A part of this time is waiting time for seed electrons. Electrons can be provided by field detachment from negative ions. This effect can be important when ions of a previous discharge are still present. It occurs in the *self-repetitive* DC corona and it causes the streamer to choose the path of its predecessor. In most applications of pulsed corona it does not occur due to the low repetition rate (~100 Hz). [VEL01]

In corona discharges at relatively low voltages the discharge stops itself due to the build up of space charge near the sharp electrode. This space charge then disappears due to diffusion and recombination and a new discharge pulse appears. This is the self-repetitive corona and it occurs in the positive and negative case [VEL01]. For negative pulsed corona, the current is due to electron motion and occurs in discrete pulses with very little current flowing between pulses and the corresponding light pulses can occur with a 1 ms period [MOR97b]. According to Raizer [RAI87], the pulse repetition rate reaches 10^4 Hz if the corona is on the anode, and 10^6 Hz if it is on the cathode.

In addition to the slow self repetition of coronas, the glow of a positive corona also pulses rapidly (~ 1 MHz) over a wide range of voltages. Morrow [MOR97a] reported that the entire area of the anode glow does not pulse simultaneously and that very few seed electrons are needed in order to sustain a regular pulsing glow. According to Morrow, detachment processes (negative ions) are crucial and positive glow corona does not occur in pure non-attaching gases such as pure nitrogen and pure argon.

Negative tip

The pulses follow a very regular manner. They are known as *Trichel pulses* (studied by G.W. Trichel in 1938). As the voltage is increased, the pulses disappear and a steady-state corona is sustained until the spark breakdown of the discharge gap.

Trichel pulses are *not observed* in the electropositive gases N_2 and Ar (N is electronegative). Air is different: once an electron is far from the point, it finds itself in a weaker field and becomes attached to a molecule. The space charge of negative ions weakens the field of the point, the multiplication of avalanches is suppressed, and the current decays. Fig. 17 shows examples of single Trichel pulses.

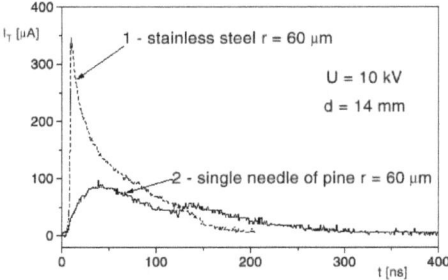

Fig. 17   Oscillograms of negative corona current pulse wave forms taken in atmospheric air and $V = 10$ kV using stainless steel and pine needle cathodes. [AUB99]

Loeb [LOL52] attempted to explain the pulses as follows: electrons, ejected from the cathode surface, create an avalanche near the cathode, causing an exponential or even more rapid rise of the current. The positive ions thus created at some distance from the cathode remain practically stationary and lead to a strong formation of positive space charge that rapidly increases the initial ionization. The electrons proceed out into the gap beyond the positive space charge. They are transformed into negative ions by attachment and thus build up a slow-moving cloud of negative-ion space charge. The positive-ion movement shortens the effective high field, although it increases its strength. This action, together with the influence of the negative-ion space charge, reduces the effective field near the cathode and leads to the discharge being quenched. Then the discharge remains extinguished until the negative space charge has dissipated.

Morrow [MOR97b] simulated the development of single current pulses and found a regular pattern of "saw-tooth"-shaped current peaks with a period of about 1 µs with a DC level. The frequency was proportional to the ion mobility and varied almost linear with the applied voltage.

Positive tip

Held [HEL97] points out a state with recurring transient current impulses that is superimposed to the direct current at the beginning of a glow discharge also in N_2, analogue to results in air. The pulses he observed had a much lower frequency than that found for Trichel pulses.

Patsch [Hoof, 1997 #275] describes discharge oscillations in inhomogeneous fields as follows: The charge carriers of the tiny local avalanche—that starts when the local electric field exceeds the initiation field E_i—produce an additional electric field ΔE that superimposes on the field E_a generated from the external voltage. The local breakdown stops as soon as the resulting local electric field is too small to maintain the avalanche mechanism, i.e. becomes lower than the extinction field E_x.

4 The DC glow discharge

The discontinuity of discharge processes is the result of space charges that periodically suppress the generation and continuation of the tiny avalanches. As a consequence of the high mobility of charge carriers in gases in a glow discharge, the repetition rate of discharges is very high with time intervals in the region of a few tens of nanoseconds (~10^8 Hz). Hence in a bigger time scale glow discharges appear to be continuous.

Comparison to our observations

In our microsystems (type Plan) we measured repetition rates of oscillating discharges. In general, the frequency increased with the gas pressure. There was no clear dependency found regarding electrode distance. Depending on the circuit ballast resistance R_b and applied voltage V_a. With R_b = 100 MΩ and 420 V < V_a < 600 V, the rate was 300 to 3500 Hz. With R_b = 1 GΩ, the rate was 30 to 90 Hz (see §5.3.3 on page 84). In earlier experiments with wire ends as electrodes and R_b = 1 MΩ repetition rates lay between 5×10^4 and 1×10^6 Hz. Raizer does not mention a resistance dependency. A comparison of the repetition rates he states with the ones we measured shows that we may have measured both anode and cathode corona, except for measurements with high R_b. Also from the geometry of our devices coronas are possible, but would at low applied voltages preferably appear on the cathode. Unfortunately we have found no publication of work on (corona) discharges where both electrodes are tips for comparison. We have not observed oscillations that reached such extremely high repetition rates and short pulse lengths typical for Trichel pulses.

4.2.3 Current oscillations in dark discharges

Current oscillations similar to those well known oscillations in corona discharges have also been observed in Townsend dark discharges in inhomogeneous fields. These are well described by Ercilbengoa et al. [ERC01]. According to Ercilbengoa the dark discharge at atmospheric pressure was formerly called *Hermstein's glow*. A mechanism involving *double-layer* formation and fluctuation is proposed to explain the discharge behaviour in the low-current domain where the applied voltage is not sufficient to cause the transition to a classical glow discharge. In an inhomogeneous field, the dark discharge presents a positive value of dV/dI. The glow regime displays recurrent impulses with current waveforms similar to the pre-breakdown streamers (and sometimes called *pseudo-streamers*). The oscillations described by Ercilbengoa remain present between the high repetitive current impulses (the frequency of which is much lower) which characterize the "classical" glow.

With increasing applied voltage, oscillations in the kHz to MHz frequency range are observed, superimposed onto the DC component, and denoted as α *oscillations*. The α oscillations appear on the ascending part of the *V-I* characteristic curve and are quasi-sinusoidal at low currents (on the order of 10 µA). Their amplitude and waveform are very sensitive to the gap voltage and, generally, they gradually transform into a more impulse-shaped waveform, until they suddenly give birth to high and sharp impulses characterizing the glow (denoted as γ *impulses* in previous studies). However, for N_2 above 3.3 kPa and also for dry air, an intermediate system of so-called α' *impulses* appears just before the breakdown and coexists with α oscillations. These impulses, which coincide with the oscillations described by Held and mentioned earlier, the amplitude of which is more than ten times the DC component, present a characteristic waveform: short rise time and slow decrease. They occur rather randomly among the oscillations, which they interrupt during a few periods. At the transition to glow they are embedded as a "base" for the γ impulses.

Double layer

The double layer consists of two opposite space charge regions that have a current limiting effect, which inhibits transition to normal glow due to a slight increase of the voltage. The double layer operates like a load resistor: electrical energy is converted into kinetic energy and thus by collisions into thermal energy (this is illustrated by the positive slope dV/dI of the characteristic curve). Under these conditions the current flow is stabilized and a steady-state DC discharge is obtained. Due to the relatively wide spatial extension of the double layer, the low-field region in the vicinity of the electrodes may be very narrow when d is less than 1 cm. In such a case the double layer has not enough space to establish, the influence of the cathode is perceptible up to the anode and the transition to glow is direct. The lower limit on the gap length, beneath which no oscillatory system can be maintained, is a few millimetres, depending on the pressure and on the tip curvature.

5 Micro gas discharge devices

We designed and fabricated a series of micro gas discharge devices. Various characteristics of these micro ionizers were measured at different pressures with several gases and for different gap distances. Most of the experiments were performed in a vacuum system that we built for that purpose. We obtained stable DC plasmas in micro machined electrode gaps from 1 to 50 µm width, at pressures up to and slightly above 100 kPa in various gases. With planar 3 µm gaps, stable glow was achieved at atmospheric pressure in Ar and N_2 respectively. With bulk Si devices stable glow was also obtained in larger gaps and in laboratory air after activation of the micro electrodes.

5.1 Experimental setup

5.1.1 Micro electrode designs and fabrication

The electrode design comprises the materials used, and the electrode and substrate geometry. Important geometrical parameters determining the discharge characteristics between two electrodes are the electrode distance and shape. The distance goes into the Paschen (-like) equation, while the geometry, including surface topography, influences the field distribution and amplification. The choice of materials was to a large extent depending on the materials commonly used in our cleanroom and for the processing of which standard processes existed or could be developed with a reasonable effort. The geometries also were influenced by the possibilities given by the micro processing techniques applied. A multitude of electrode combinations was designed. The following table gives an overview of the processes, including mask designs, that have been fabricated and used. More detailed process descriptions can be found in the appendix.

Table 12 Overview of ionizer processes.

Process name	Substrate(s)	Processes used	Comments
Planar (*Plan*)	Fused silica	Deep dry etch of substrate below Cr/Pt	Most of the experiments were done with this design. Lifetime: several hours in low *p* Ar.
Insulated Planar (*I-Plan*)	Fused silica	Cr/Au on Pyrex, SiO$_2$ insulation layer on top	First complete analyzer designs. Short lifetime due to rapid evaporation of electrodes and development of short circuits between the electrodes.
Bulk Si (*Si-Bulk*)	Si + Pyrex	DRIE of Si, anodic bonding to Pyrex	Highest current, best long term performance. Lifetime: > 14h in 100 kPa N_2.

Thin Pt electrodes on deep etched fused silica—Plan

The structures of Plan were the first to be used for microplasma experiments. The electrodes consist of a 300 nm thin layer of Pt on a fused silica substrate. The substrate was dry etched to a depth of 5 to 10 µm to achieve electrodes that are elevated above the substrate. Like this, the field strength of the edges exposed to the gas is higher, and the conduction through surface contamination is reduced because of the longer leakage path. These electrodes are illustrated in Fig. 18, a SEM image of the same structures is shown in Fig. 19.

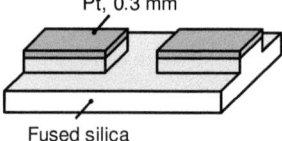

Fig. 18 Schematic drawing of Plan microionizer structures. A layer of 300 nm of Pt is structured on a fused silica substrate. The substrate is subsequently dry etched by 5-10 µm to achieve elevated electrodes.

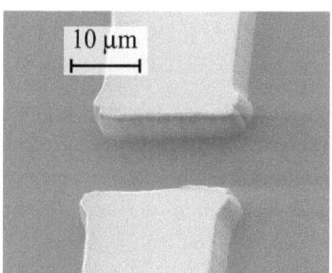

Fig. 19 SEM image of Plan electrode structures, as seen from an angle of about 30°.

Insulated thin Au electrodes on fused silica—I-Plan

In the experiments with Plan electrodes it was found, that oscillations of the microplasma with high amplitude and low frequency could be reduced by partly insulating the cathode with photoresist, as shown in Fig. 50. In the design of I-Plan the cathodes were insulated with a layer of SiO_2. Au was used for the electrodes because the dry etch that is necessary for Pt could not be applied in this process that was also used for other structures. Fig. 20 shows a sketch of the I-Plan electrodes. In Fig. 21 a strong underetch of the SiO_2 insulation layer that occurred during the wet etch in RF-solution can be seen.

5.1 Experimental setup

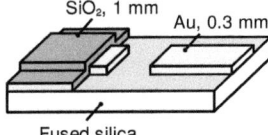

Fig. 20 Schematic drawing of I-Plan microionizer structures. A layer of 300 nm of Au is structured on a fused silica substrate. SiO$_2$ serves as an insulation layer to avoid spreading of a glow discharge over the cathode surface.

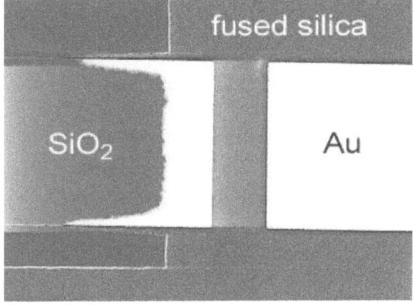

Fig. 21 SEM image of I-Plan electrode structures, top view. The SiO$_2$ layer is strongly underetched along the edges of the Au electrodes. The bright area between the electrodes is an artefact from the SEM.

The design of I-Plan did not allow for deep etching of the substrate. Therefore material that was evaporated from the electrodes and deposited on the substrate between the electrodes caused a short circuit within minutes, see Fig. 33 and Fig. 34 on page 73. Also, the low thickness of the electrodes led to rapid destruction of the electrodes due to sparking. Therefore extensive experiments were not possible with I-Plan, and the samples were mainly used for breakdown measurements.

Bulk Si electrodes bonded to Pyrex—Si-Bulk

To achieve thick electrodes that are well detached from the substrate, we etched 50 µm deep trenches into a <100> Si substrate by dry etching, then liberated these electrodes by deep anisotropic wet etching from the backside. The substrates were then anodically bonded to Pyrex wafers and diced. Fig. 22 and Fig. 23 illustrate the result. Thanks to their thickness, these electrodes showed a long lifetime even during arc discharge.

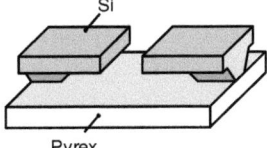

Fig. 22 Schematic drawing of Si-Bulk microionizer structures. Thick electrodes are etched into a Si substrate that is then bonded to a pyrex wafer.

5 Micro gas discharge devices

Fig. 23 SEM image of Si-Bulk electrode structures, seen from an angle of about 30°. The upper left electrode is stretched sideways.

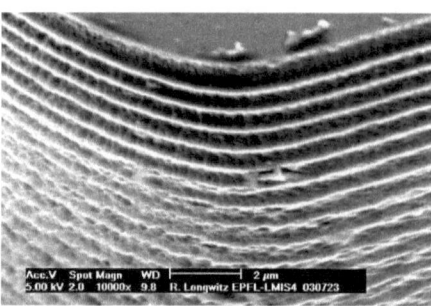

Fig. 24 Enlargement of the electrode sidewall of the SEM of Fig. 23. The surface is rough and comprises sharp tips that may serve as electron sources when a high voltage is applied.

5.1.2 Setup for micro ionizer experiments

Most of the measurements were done in a vacuum system where we could control the gases and pressures. With the electrode structures Si-Bulk, it was possible to do experiments in laboratory air in which the earlier designs had a too short lifetime. Some of the ionizers were eventually tested in our miniature ion mobility system that is covered in the next chapter.

The vacuum system comprised a window for the observation of discharges, a chip holder that allowed to contact up to eight electrode pairs, and a multi-feedthrough for easy plugging of the connected electrode pairs. A schematic drawing of the system can be found in the appendix. The electric circuit that was used in the experiments is shown in Fig. 25. The applied voltage and the circuit current where both displayed at the same time on a two channel oscilloscope. Fig. 26 shows a photograph of the window in the vacuum system through which the ionizer could be observed. Discharge experiments were done in a grey room (reduced dust content in the air) and the laboratory was darkened during the experiments.

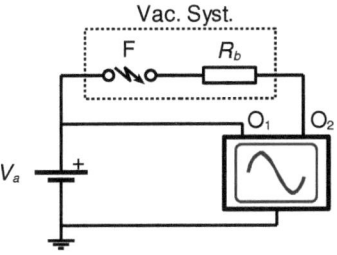 

Fig. 25 Electrical scheme of the circuit used for discharge experiments. V_a: High voltage source (DC), F: discharge gap, R_b: Ballast resistance, O_1: Oscilloscope channel for measurement of input voltage, O_2: Osc. channel for measurement of circuit current (i.e. voltage over 1 MΩ internal resistance).

Fig. 26 Window of the vacuum system used for the measurements under controlled atmosphere. Inside, the chip holder holding an I-Plan chip can be seen. During experiments, the chip was observed with a microscope.

5.2 Discharge experiments with micro electrodes

In our experiments, we varied the major parameters and observed their effects on the discharge: Electrode design, electrode distance d, gas type (air, N_2, Ar, CO_2), gas pressure p, ballast resistor R_b. The leakage current was negligible (resistance of several TΩ). The first design that allowed us to observe microplasma in the controlled atmosphere of our vacuum system was Plan, thin Pt electrodes on deep etched fused silica. These electrodes were sensitive to sparking which occurred in almost all experiments and their lifetime was limited, even at low pressure in a noble gas like Ar. Because of edge erosion, the electrode distance could not be given exactly during the experiments, except for the measurements of the breakdown voltage of unused electrodes.

5.2.1 Breakdown voltage

In breakdown experiments, the voltage at which a gap breaks down under certain conditions is determined. What is needed for such experiments are gaps of the desired spacing. We recorded breakdown voltages during glow experiments and did additional breakdown experiments with our micro devices, where the electrode distances are defined by mask and process. Similar experiments have been conducted before [ONO00, ONO00b]. Another approach is to use discrete electrodes, one of which can be displaced with sub-micron resolution [DHA00, DHA94, TOR99b]. Such a setup is quite complex and delicate, surface finish and cleanliness have a strong influence on the results.

The advantage is, that nearly plane parallel electrodes can be used, approximated by two large spheres, for example.

The fields in our devices are inhomogeneous, a deviation from Paschen's law therefore had to be expected even with large electrode gaps, as explained before. Still, we found agreements in the general behaviour. According to the Paschen equation the breakdown field strength E_b is approximately constant in gas for $pd \geq 10$ Pa×m. Davies [DAV73] observed that the local cathode electric field for breakdown E_b in vacuum is approximately constant for gaps with $d > 25$ µm. In our own measurements with small gaps ($d \leq 50$ µm) E_b tended towards a constant value at greater d, almost independent of the gas pressure. The breakdown voltage V_b varied greatly, as can be seen in the collection of breakdown measurements shown in Fig. 27 (300 V < V_b < 750 V). The average remained about constant over the whole range of pd, neither p nor d had a strong influence on the breakdown voltage. The breakdown field strength E_b is therefore increasing towards small gaps with $1/d$, see Fig. 28.

For the calculation of E_b equation (3.18) must be used:

$$E_b = \beta \cdot \frac{V_b}{d} \qquad (3.18)$$

Fig. 28 shows values for $\beta = 1$ (parallel plates, denoted E_{b1}) and $\beta \approx 10$. A β of about 10 would result, for example, from a protrusion of 100 nm height and a radius of 5 nm, calculated using the results described in section 3.2.1 on page 36. The edge effect was not taken into account. As stated in §3.2.1, the threshold for emission of electrons is about 3×10^9 V/m. The highest E_{b1} we measured was 4.3×10^8 V/m with $d \approx 1$ µm. Even with a field amplification factor due to local surface protrusions as low as 10, we are already in the range of appreciable electron emission that supports the initiation of breakdown.

5.2 Discharge experiments with micro electrodes

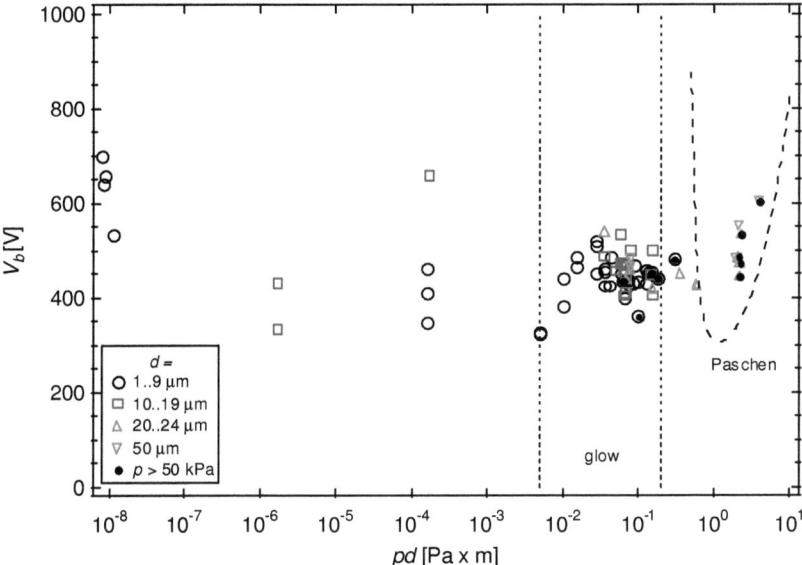

Fig. 27 Breakdown voltages V_b over pd values of breakdown measurements in N_2. Same data as in Fig. 28, 1 µm $\leq d \leq$ 50 µm, see legend. Points measured at $p > 50$ kPa are marked with a dot. The dotted vertical bars denote the approximate range in which glow has been observed. The dashed line is the Paschen curve as calculated for N_2.

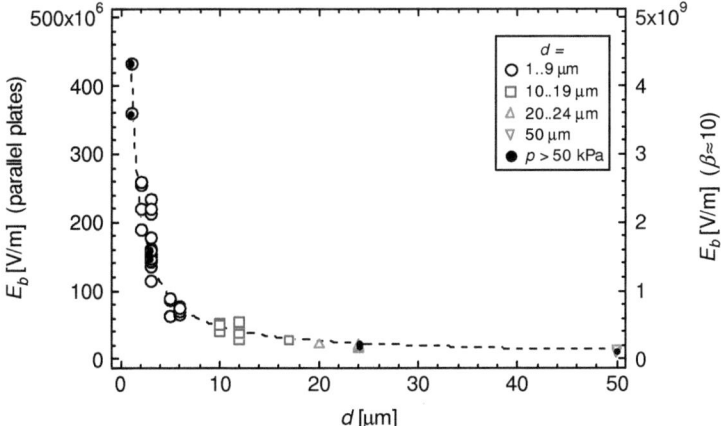

Fig. 28 Breakdown field strength at various gap distances and pressures in N_2 measured with various ionizer designs. Same data points for both graphs. E_b vs. d; $2.7 \times 10^{-3} \leq p \leq 1.7 \times 10^5$ Pa. Points measured at $p > 50$ kPa are marked with a dot. The fit through the measured data (dashed line) tends towards $E_b = 5 \times 10^6 \pm 3 \times 10^6$ V/m for $d \rightarrow \infty$. 3×10^6 V/m is generally regarded as the minimum breakdown field strength in air.

The lowest measured E_{b1} was 9.3×10⁶ V/m at $d \approx 50$ µm. Here, a β of at least 300 would be necessary to allow for similar emission as in the case of $d = 1$ µm. This we cannot exclude, but still we interpret the distance where the measured E_b starts to increase quickly as the transition to field strength dependent breakdown where field emission becomes important. Gas type and pressure become less important in this range, as also explained in §3.2.1. Our measurements do therefore correspond well to the theory, but differ considerably from Torres' [TOR99b] results, see Fig. 29, while agreeing better with Germes' [GER59] measurements down to 1 µm gaps, see Fig. 30.

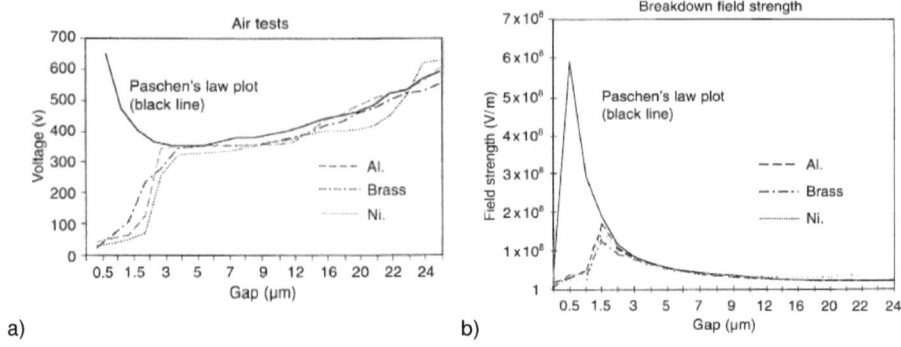

Fig. 29 Graphs showing the results of the breakdown measurements of Torres and Dhariwal [TOR99b] with two polished metal spheres as electrodes.

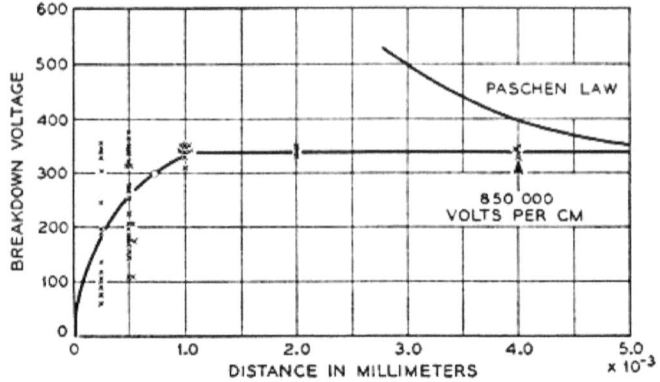

Fig. 30 Graph showing results of Germer [GER59] for silver electrodes in air which were well polished at the start of the tests.

In Torres experiments, the breakdown voltage as well as the field strength tend towards zero for separations of less than 1.5-2 µm, while Germer found an almost constant V_b down to 1 µm. Below

1 μm, V_b in Germers measurements vary greatly, the highest measured V_b still being at and even above 340 V, as at greater separations. We consider the highest measured breakdown voltages as the more reliable ones, because any protrusions, particles or other contaminations that have an influence on V_b will rather reduce it, especially at low d. Germer states that breakdown in his measurements in the range from 1 to 4 μm was *obviously* air breakdown over a path longer than the measured electrode separation. Also, Torres' and our own discharges might have occurred over a longer path than the shortest between the electrodes. This seems likely with Torres' ball shaped electrodes. It seems less likely for our structures: The way our electrodes are shaped, the field strength reduces quickly away from the gap, as shown in Fig. 31. Breakdown will therefore likely not occur on a longer path than the gap distance. In later experiments we observed arcing over far distances from the anode edge to some point on the cathode, but the arcing distance reduced with the applied voltage down to the actual gap distance.

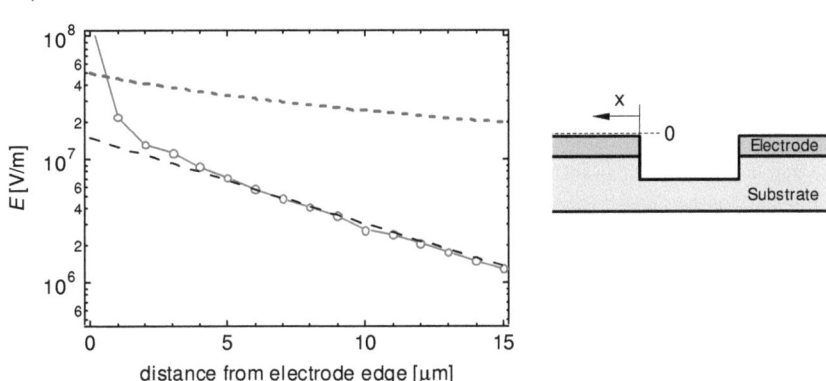

Fig. 31 Field strength above the cathode over distance from the electrode edge, simulated for electrodes of type Plan, d = 10 μm, V = 500 V. The dotted line shows the field strength for parallel plates. Dashed line: fit with $a \times \exp(-b \times x)$ where $a = 1.5 \times 10^7$, $b = 0.16$.

Since the gap has a certain width, discharge could also occur from a protrusion on one electrode edge to a protrusion on the other electrode edge, where these protrusions do not directly face each other. Then also the discharge distance would not equal the gap distance d. The edges on our microstructures are so rough, that there are small protrusions all along the edges. Then we expect breakdown to occur between the points where the field strength is highest i.e. between two protrusions that are not much displaced. In a certain range of pd, not far left of the Paschen minimum, a breakdown could still decide to take a longer path, but likely not *much* longer. Our measurements have not given evidence of such effects. However, there remains an uncertainty about the actual breakdown path length.

5 Micro gas discharge devices

5.2.2 Glow range

There is a range of pressures and voltages where glow is possible in a given small electrode gap. When a glow has started, there may still be sparks appearing, usually from the anode tip to a point on the cathode edge beyond the glow area. When increasing the pressure, the edge of the glow and sparks move towards the gap, until the glow stops. Increasing V_a will then only lead to more sparking. When decreasing p during a stable glow, the glow extends until it eventually stops without sparking. No stable glow was achieved in 1 μm gaps, with or without insulated cathodes. Probably the pressures used, less than or only slightly above 100 kPa, were not high enough. Möller [MOE99a] reports a glow at 500 kPa in a 5 μm gap in He.

The range of pd where stable glow in N_2 was observed with planar electrodes is approximately from 0.005 to 0.2 Pa×m in gaps from 3 μm to 50 μm. This range is indicated in Fig. 27. The glow range we found for Ar was 0.02 to 0.14 Pa×m. It must be taken into account though, that neither pressure at low range, nor the electrode distance was measured accurately.

The smaller the gap size d, the greater the necessary pressure p in order that the fall region of the cathode layer, vital for a glow discharge, can be smaller than the electrode distance. If the mean free path between ionizing collisions of electrons with gas molecules is too long compared to the electrode gap, no avalanche can form to enable a self-sustaining glow discharge. Instead sparking will result from electron emission when it becomes so high that electrode material evaporates [LAT81]. The necessary gap potential for sparking in such a case is much higher than for a normal glow discharge, therefore a glow turned off without sparking when we reduced the pressure during a glow.

Above the maximum pd where glow occurs, the mean free path is so low that electrons cannot reach sufficient energy between two collisions unless the gap potential is so high that streamers, i.e. sparks, can form. In our experiments, when a glow discharge turned off when increasing the pressure, sparks appeared. When a glow turns off, the gap voltage increases because of the increased gap resistance. This may account for the immediate appearance of sparks. We did not endeavour to *calculate* the glow range.

Working at a high pressure, i.e. close to the maximum pd for glow, has the advantage that a greater number of collisions reduces the energy of positive ions in a discharge and thus sputtering on the cathode is reduced.

5.2.3 Effects of discharges on electrodes

Depending on the electrode design and experimental conditions we observed a variety of effects on the electrodes during discharge. Thin metal film electrodes generally had a short lifetime and were especially sensitive to sparking. The thick electrodes of Si-Bulk showed other effects: material was deposited on the surfaces, and whiskers grew on the edges.

Electrode damage

The following figures show the destructive effects of discharges on thin film electrodes. Electrodes and electrode gaps deteriorate in three ways: melting and evaporation, deposition of evaporated material that can cause short circuits, and sputtering of electrode material. Fig. 32 shows the effect of sparking on Pt electrodes in Ar. In Fig. 33 you can see the melted edge of an Au electrode and Au deposited on the substrate. Here, the gap failed by a short circuit. The evaporated metal can spread far beyond the gap as is demonstrated in Fig. 34. Please note that the gap distances d given are the nominal distances that increased when the electrodes deteriorated, and the values for the pressure of Ar are approximate, because the gauge was not calibrated for Ar.

Fig. 32 SEM image of Plan electrodes after discharge experiments. The electrode to the left was the cathode. The Pt layer on top of the fused silica substrate has partly melted and evaporated on both electrodes.

$R_b = 100$ MΩ, $d = 10$ µm, gas: Ar at 1 kPa. $V_a > 500$ V.

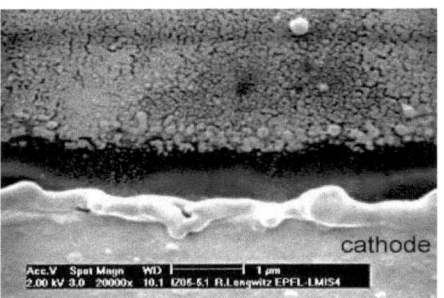

Fig. 33 SEM image of an I-Plan cathode edge after discharge experiments. On the bottom the partly melted and evaporated edge of the Au electrode can be seen. The rough surface on the upper half of the picture is Au that has deposited between the electrodes, causing a short circuit.

$R_b = 100$ MΩ, $d = 3$ µm, gas: Ar at < 1 kPa. $V_a \leq 743$ V.

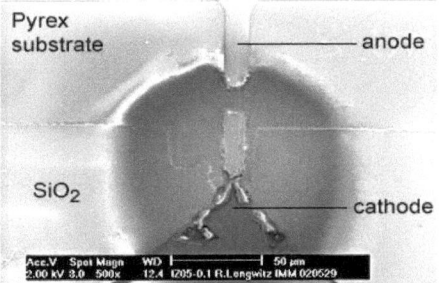

Fig. 34 SEM image of I-Plan electrodes after discharge experiments. The dark circular area shows where Au, evaporated from the electrodes, has deposited during 5 min. of stable glow. The surface around this circle appears brighter because of charging effects in the SEM.

$R_b = 100$ MΩ, $d = 6$ µm, gas: Ar at 10 kPa. $V_a \leq 500$ V.

Deposits and whiskers on electrodes

The former experiments led to the conclusion, that the electrodes must be significantly thicker in order to achieve longer lifetimes even in a more aggressive environment, like air at atmospheric pressure. Such robust electrodes were achieved with Si-Bulk, where the electrodes consist of thick Si bonded to Pyrex. In order to avoid a short circuit by deposited electrode material and to minimise the influence of the insulating substrate on the discharge, the electrodes are suspended above the substrate. These structures enabled experiments outside the vacuum system that had not been possible with the former designs. They were operated for several hours without significant deterioration. Not only new phenomena of the discharges themselves, but also very different surface effects were observed, as shown in Fig. 35 to Fig. 37.

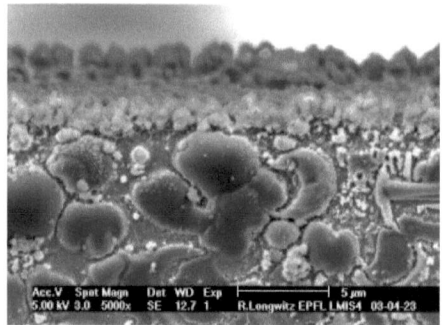

Fig. 35 SEM image of a Si-Bulk cathode edge after discharge experiments, top view. Here, a contaminant has deposited not only on the sidewalls, but also on the other surfaces. This occurs, when the glow expands beyond the gap over the electrodes at high voltages or at low pressures.

R_b = 100 MΩ, d = 36.5 µm, gas: Ar at 26.3 kPa, pd = 0.95 Pa×m. Continuous glow during 1h34m at up to 420 V.

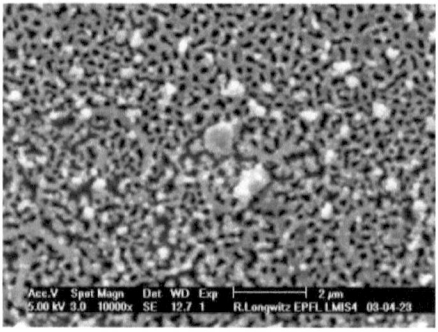

Fig. 36 An Si-Bulk cathode surface after discharges, top view. At high pressure, pores formed on the cathode surface.

R_b = 100 MΩ, d = 36.5 µm, gas: Ar at 110 kPa, pd = 4.0 Pa×m. Continuous glow during 1h at up to 1300 V.

Fig. 37 Another Si-Bulk cathode surface after discharges in air.

R_b = 100 MΩ, d = 16 µm, gas: air at 100 kPa, pd = 1.6 Pa×m. Continuous glow during 4h6m at up to 600 V.

Electrode activation

Activation (or *conditioning*) generally leads to a roughening of the surface, if the surface is not destroyed altogether. We observed activation on microelectrodes as well as on miniature electrodes that we used for ion mobility experiments, see §6.1.

Activation in miniature electrodes

We observed that the glow between unused electrodes starts at relatively high voltages and is very noisy at first. After glowing for about 1 min in air, the glow becomes more stable, and restarting of a glow with such "activated" or "conditioned" electrodes starts at a much lower voltage (about 100 V lower). The activation effect seems to be due to a growth of a carbon layer with sharp tips (whiskers) on the electrode surfaces. We found that the activation develops much more rapid with an admixture of isopropanol (($CH_3)_2CHOH$) or acetone (C_3H_6O) in N_2 than in pure N_2. In N_2, the discharge continued for 40 s to 1 min, while admixtures of isopropanol or acetone lead to a short circuit within a shorter time, down to 15 s.

The activation process also depends on the electrode materials. Activation of tungsten needles with a steel counter electrode was successful, while aluminium as counter electrode material did not lead to activation. We assume that carbon in the electrode material supports needle growth on the surfaces. An example of an activated tungsten tip is given in Fig. 38.

5 Micro gas discharge devices

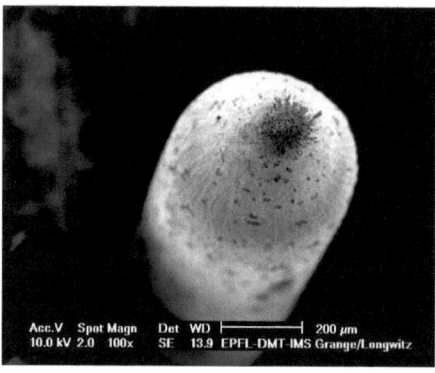 

Fig. 38 Activated tungsten needle. Left: Needle tip with structures, likely carbon, grown on the original tip. Right: Detail of the grown structures.

Activation of micro electrodes

An example of activation is shown in Fig. 39, where whiskers have grown along the electrode edge in a glow discharge. The tips that have grown on the anode in Fig. 40 are almost touching the cathode. This would cause a short circuit, but not necessarily destroy the ionizer, because the current can evaporate the connecting "filaments" as supposedly has happened to the tips shown in Fig. 41.

a) 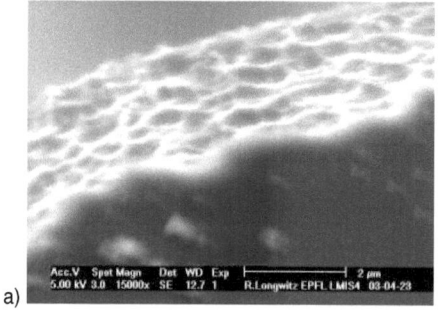 b)

Fig. 39 SEM images of a Si-Bulk cathode edge after discharge experiments, top view. The dark areas are the top surfaces of the electrode, the brighter areas are the "vertical" sidewalls. a) Due to the round shape, the field strength in the gap reduces towards the electrode corners. No trace of discharge effects can be seen. b) The same edge as in a), near the electrode centre. Whiskers have grown on the surface. Breakdown occurred at 355 V when the electrode was first used. After a short time (< 1 min) of stable glow the same gap broke down at 303 V—an activation effect.
R_b = 100 MΩ, d = 36.5 μm, gas: Ar at p = 25.6 kPa, pd = 0.93 Pa×m. Continuous glow during 1h10m at up to 870 V.

5.2 Discharge experiments with micro electrodes

Fig. 40 Activation tips from the anode almost touching the cathode of Si-Bulk electrodes. $R_b = 10$ MΩ, $d = 8$ μm, gas: dry air at 100 kPa, $pd = 0.8$ Pa×m. Continuous glow during 4h41m at up to 470 V.

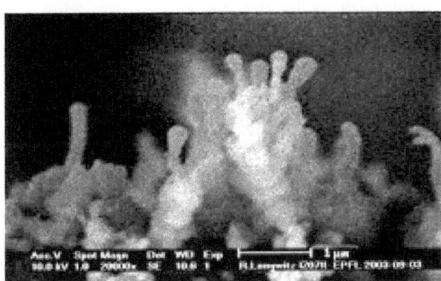

Fig. 41 Rounded activation tips on the same sample as in Fig. 40.

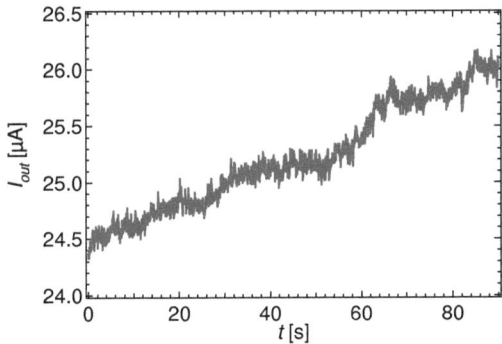

Fig. 42 A current increase during the initial phase of a glow discharge indicates the growth of protrusions.

A rougher surface leads to higher field strength near the tips of the roughnesses (whiskers etc.), which increases ionization or electron emission near these tips. Discharge is thus facilitated and the current increases, as can be seen in Fig. 42.

5.2.4 Special electrode profiles: Rogowski, Bruce and our own

In some cases, especially when sparking from electrode edges should be avoided, it is desired to avoid high electric fields on the edges. Then the edges need to be rounded. The solution according to Rogowski is to construct electrodes that follow an equipotential surface at some distance from an imaginary flat plate electrode with sharp edges. The equipotential surface is found by means of a finite element simulation of the field between two parallel flat planes. For details, see Cobine [COB58].

Bruce's version of electrode shapes only approximate an ideal uniform field. The Bruce profile is a figure of revolution, starting with a flat plane in the centre, with a sine curve used as a transition to a circular section at the edge. The idea is to have a large area of uniform field (2 flat plates) with a gradually decreasing radius of curvature to the edge. For details see [CRA54].

We approximated Bruce's shape by a quarter of an ellipse, which is somewhat easier to draw. As a standard shape, We chose a 2:1 ellipse where the smaller radius equals the gap size. This radius gives an electrode shape that approximates shapes of formerly sharp edges rounded by discharges, as can be seen in Fig. 43. Of course, because the electrode is a thin layer, the layer edge already causes strong field enhancement. Still it is important to avoid additional field concentrations at the corners. An FEM simulation of the field between two corners, one of which is rounded in the described way is shown in Fig. 44.

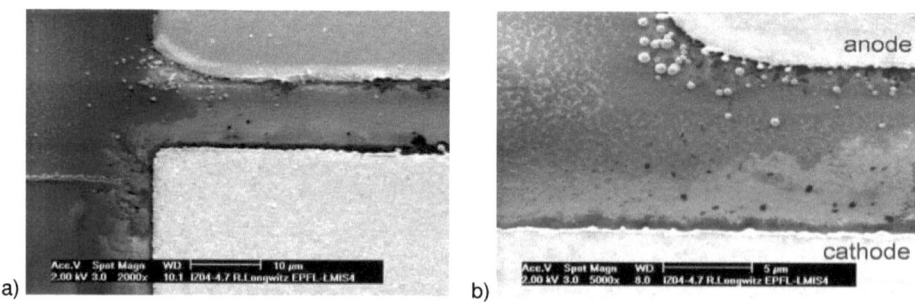

Fig. 43 SEM image of I-Plan electrode edges after discharge experiments. The "corners" on the upper edge in a) and of the edge in b) were right angle corners before the discharges. They were rounded by the discharges, because the electrical field strength was higher at these points then in the rest of the gap. In more recent designs, such rounded shapes were used for all electrode corners.

5.3 Simulation of glow discharge oscillations

Fig. 44 A simulation of the field strength between the electrodes when a 2:1 quarter ellipse shape is used for one of the electrodes. The field strength is still slightly elevated where the curvature begins in the gap, but then reduces smoothly. The colour gradient is linear.

5.3 Simulation of glow discharge oscillations

In our experiments with glow discharge in micro gaps, we observed oscillations in the circuit current. From these observations and physical considerations we deduced an electrical equivalent circuit. Parameters were extracted from measured oscillations and directly attributed to the model circuit components. We found that the simulated results were in good agreement with our observations of oscillations. The circuit then allowed for a mathematical analysis of the observed discharges and serves the understanding of phenomena and the prediction of effects of the experimental parameters on discharges. It enabled us to explain the improvement of the stability of a glow discharge by partially insulating the cathode.

The oscillations we describe here phenomenologically correspond to intermittent positive corona discharges as explained by Raizer [RAI87], for example. Simulations of such discharges were presented by Morrow [MOR97a] and Akishev et al. [AKI99]. They applied continuity equations and the Poisson equation to simulate oscillations of positive coronas. Both dealt with electrode gaps on the order of 1 cm and found oscillation frequencies > 100 kHz, much higher than the frequencies on the order of 1 kHz found in our micro discharge devices. Equivalent circuits for the simulation of partial discharges have first been developed by Gemant and Philippoff [GEM32] and used extensively especially in the early 1990s. To our knowledge, we are the first to apply this method to model the oscillations of a glow discharge.

5 Micro gas discharge devices

5.3.1 Description of an oscillation cycle

Fig. 45 shows a scheme of our experimental setup. We give a simplified description of the phenomena occurring in the circuit during an oscillation cycle that is then used as the basis for our model. In an experiment, the applied voltage V_a is slowly increased from 0 V. The gap resistance is infinite, so the whole voltage drops over the gap and there is virtually no current. When the breakdown voltage V_b is reached, the gap breaks down, i.e. an avalanche develops, molecules are ionized and space charge builds up in the gap. When a current flows through the gap, the gap voltage V_g decreases until it is not anymore sufficient to keep the discharge going, while the space charge partly counters the applied field. This voltage we call the extinction voltage V_x. Here the avalanche dies out, but the current continues due to the charges that are still in the gap. The current decreases with the number of neutralized charges, while the gap voltage and the field recover until the next breakdown and the cycle repeats. After the initial breakdown, further breakdowns occur at a lower voltage because of the ions that are already present in the gap. These have an effect similar to an external ion source as mentioned by Held et al. [HEL97]. We denote this reduced breakdown voltage with V_b'.

5.3.2 Model circuit for oscillating discharges

Fig. 45 shows the physical circuit that was used for the discharge experiments and that corresponds to the circuit in Fig. 25. For our simulations we applied the circuit shown in Fig. 46.

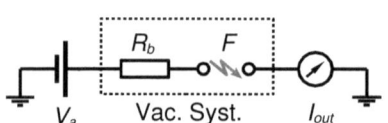

Fig. 45 Electrical scheme of the circuit used for discharge experiments. V_a: applied voltage, F: discharge gap, R_b: ballast resistance, I_{out}: current, measured with an oscilloscope. (1 MΩ impedance).

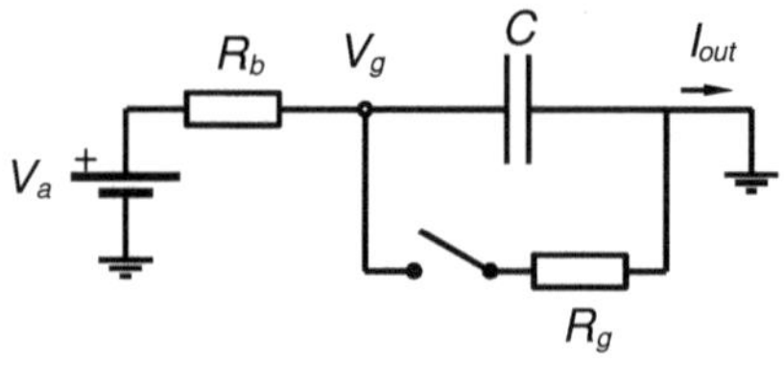

Fig. 46 Electrical scheme of our model circuit. V_g: gap voltage, C: gap capacity, R_g: gap resistance.

The main circuit resistance is represented by R_b. The gap is modelled by a switch, in line with a resistor R_g and parallel to a capacitor C. The switch closes when the gap voltage V_g increases above V_b', and opens, when V_g decreases below V_x. The capacitor stands for the capacity of the gap, which is partly made up of the geometrical capacitance of the electrodes. The other part is charge that builds up in the gap when the discharge starts and wears off when it stops: we call this the *plasma capacitance*. C and R_g can be extracted directly from the measurements.

There is one important difference between the behaviour of the real discharge gap and our model: The real gap corresponds to an empty capacitor before discharge. At the moment of breakdown this "capacitor" is loaded quickly with space charges, which slowly vanish when the avalanches cease. In the model just the inverse happens: While the capacitor charges slowly when the switch is open, it discharges quickly at "breakdown". This is a physical flaw of the model, but we thus obtain a good match for the measured and modelled output current as shall be shown.

Gap resistance

The gap resistance R_g corresponds to the specific resistance ρ introduced in §3.2.1 on page 26 plus the "contact" resistance caused by the resistance that is opposed to neutralization of electrons or ions at the electrode surfaces. During a discharge the current is mainly due to electrons. Their mobility is high and the "contact" resistance low compared to ions, therefore R_g is much lower during a discharge, than after the end of a discharge. When there are no charges in the gap, R_g is infinite. In the model we assume R_g to be infinite already when the discharge stops, signified by an open switch in the circuit.

Geometrical capacitance

To calculate the geometrical capacitance C_g we assume an electrode geometry as shown in Fig. 47.

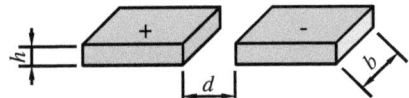

Fig. 47 Sketch of the electrode structure assumed for capacity calculations.

To find an approximate value for the geometrical capacity C_g of the electrode structures, we used the equation:

$$C_g = \frac{b \cdot h}{d} \cdot \varepsilon_0 \cdot \varepsilon_r \cdot f \qquad (5.1)$$

with electrode height h, width b, and the form factor f. The form factor is used to account for the increased capacity due to the increased field strength near the edges and the additional nonparallel area. f was found by a comparison of the capacity calculated only with the facing parallel areas to the capacity found from an FEM simulation of the whole structure. This method gives no exact values, especially when varying d from the d used in the simulation. A simulation of the capacity for a structure with $d = 3$ µm gave a factor of about 19 without substrate (as in Fig. 47), the substrate accounted for an additional factor of about 2. For the calculation we therefore chose $f = 40$ for structures Plan and I-Plan, and $f = 20$ for Si-Bulk. The relative permittivity was $\varepsilon_r = 1$. Results for various structures are given in Table 13.

Table 13 Example results for calculations of electrode capacities. The time constant τ is calculated with $\tau = R_b C_g$, where $R_b = 100$ MΩ.

Type	b [μm]	h [μm]	d [μm]	C_g [fF]	τ [μs]
Plan	20	0.3	10	0.2	0.02
I-Plan	20	0.3	10	0.2	0.02
Si-Bulk	200	40	12	118.1	11.8

These calculations show, that the capacity of the devices themselves do not contribute much to the observed oscillations, that had relaxation times close to and in the ms range. Even an uncertainty factor in the calculation of C_g in the order of 10 would not change this result.

Plasma "capacitance"

A way to calculate the capacitance effect due to space charges is to calculate the possible amount of charge that may be stored in a gap. A calculation is presented here to give one example:

Assumed parameters:
Glow volume: 3.6×10^{-16} m³ (with $d = 10$ μm, $h = 0.3$ μm, $w = 20$ μm, spread over the cathode: 20 μm, glow thickness: $2h = 0.6$ μm)
For N_2, 300 K, 10^5 Pa: 2×10^{10} molecules in volume
Charge in gap at 1 % single ionization: $Q = 3.5 \times 10^{-11}$ C
$C = Q/V, V = 500$ V $C = 70$ fF

At this ionization ratio the calculated capacity is still very small. But unlike in a real capacitor, in the gap it takes time for the ions to reach the electrodes and then to neutralize. We assume that this additional time accounts for the additional "capacitance" that leads to a time constant in the order of 1 ms that we measured.

Analysis of the model circuit

To calculate the development of the gap voltage V_g during one pulse we need two equations: one describing the circuit with an open, the other with a closed switch. We start with an uncharged capacitor and $V_a = 0$ V, the switch is open. When a voltage V_a is applied, V_g increases towards V_a according to

$$V_{g\_open}(t) = V_a \cdot (1 - e^{-\frac{t}{\tau_{open}}}) \tag{5.2}$$

where $\tau_{open} = CR_b$. When V_b' is reached, the switch closes, C discharges, and V_g decreases from V_b' towards the limit

$$V_{g\_closed}^* = V_a \frac{R_g}{R_b + R_g} \tag{5.3}$$

along the curve
$$V_{g\_closed}(t) = V^*_{g\_closed} + (V'_b - V^*_{g\_closed}) \cdot e^{-\frac{t}{\tau_{closed}}} \tag{5.4}$$

with $\tau_{closed} = C \dfrac{R_b R_g}{R_g + R_b}$.

When V_g reaches the lower limit for discharge V_x, the switch opens, and the cycle restarts, but now, since the capacitor is already charged to V_x, V_g increases towards V_a following

$$V_{g\_open}(t) = V_x + (V_a - V_x) \cdot (1 - e^{-\frac{t}{\tau_{open}}}) \tag{5.5}$$

We find the time t_{open} (recover period) for one charging process from V_x to V_b' using a simplified form of (5.5):

$$V^*_g(t) = (V_a - V_x) \cdot e^{-\frac{t}{\tau_{open}}} \tag{5.6}$$

From the condition $(V_a - V_x) \cdot e^{-\frac{t_h}{\tau_{open}}} = V_a - V_b'$ \hfill (5.7)

we conclude
$$t_{open} = -\tau_{open} \cdot \ln\left(\frac{V_a - V_b'}{V_a - V_x}\right) \approx \frac{1}{f} \tag{5.8}$$

In the same way we find

$$t_{closed} = -\tau_{closed} \cdot \ln\left(\frac{V_x - V^*_a}{V_b' - V^*_a}\right); \quad V^*_a = V_a \frac{R_g}{R_b + R_g} \tag{5.9}$$

The oscillation frequency is then

$$f = \frac{1}{t_{open} + t_{closed}} \tag{5.10}$$

If $R_g \ll R_b$, then $\tau_{closed} \ll \tau_{open}$, and we can neglect the decreasing part of V_g. Fig. 48 gives an illustration of the above equations.

5 Micro gas discharge devices

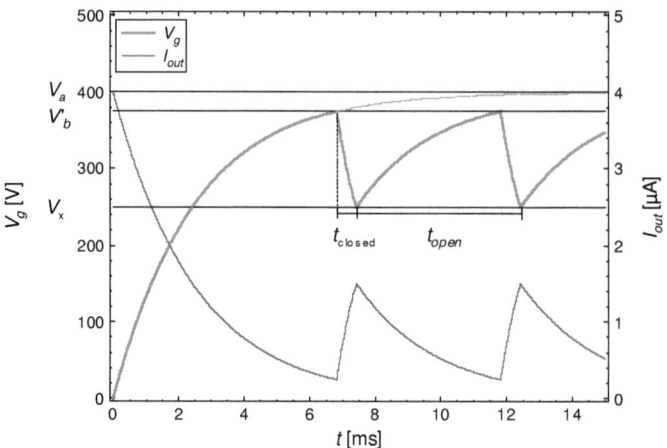

Fig. 48 Course of the gap voltage calculated for the circuit in Fig. 46. When V_a is applied at $t = 0$, the gap voltage V_g increases towards V_a, then drops to V_x at breakdown. The thin line shows the current through the circuit. Parameters similar to example A were used: $V_a = 400$ V, $V_b' = 375$ V, $V_x = 250$ V. $C = 24.6$ pF, $R_b = 100$ MΩ, $R_g = 50$ MΩ.

To model discharges, the following parameters are taken from measurements: applied voltage V_a, breakdown voltage V_b', lower limit of a discharge V_x, capacity C, time constant τ_{open}, and gap resistance R_g. C and τ_{open} are found from an exponential fit of the falling curve of I_{out}. When $t_{open} \gg t_{closed}$ (example A and B), R_g is calculated with

$$R_g = \frac{V_a - V_x}{I_{max}} - R_b \qquad (5.11)$$

For the oscillation in example C, the relationship $t_{open} \approx 2 \times t_{closed}$ from the measurement was used to calculate the capacity. R_g was then found numerically.

5.3.3 Micro discharge experiments and comparison with simulations

Experimental Setup

The measurements were done in our vacuum system in N_2 at a pressure of 10 kPa. The electrodes used for these experiments were of type Plan as shown in Fig. 18 and Fig. 19. For some of the experiments the cathode of such ionizers was partly insulated with photoresist, see Fig. 49 and Fig. 50. The gap between the electrodes was initially 10 to 12 μm, and increased by a few μm during discharges due to erosion that causes an increase in breakdown voltage. We employed a ballast resistor of 100 MΩ.

5.3 Simulation of glow discharge oscillations

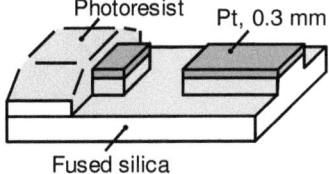

Fig. 49 Schematic drawing of Plan microelectrode structures with partly insulated cathode.

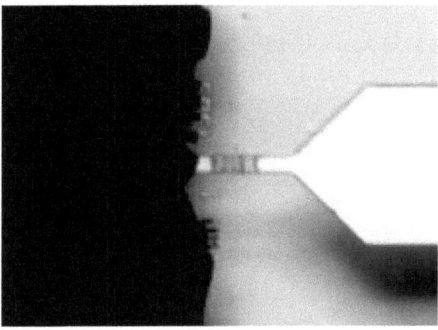

Fig. 50 Photography of Plan electrodes where the cathode was insulated with resist, applied manually. The electrodes were damaged by discharges.

Observations

A few typical measurements were chosen for examples. Since we are working with microelectrodes of a shape and gap distance that do not correspond to electrodes used for earlier experiments mentioned in the literature [ERC99, GOL03], the pulse shapes partly differ significantly from these.

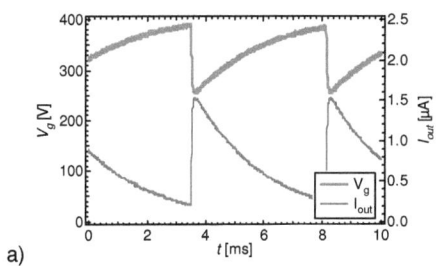
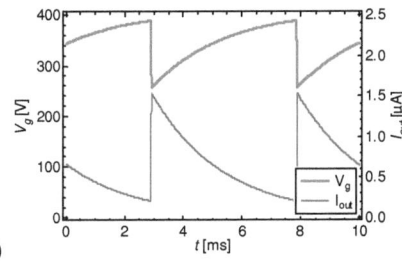

Fig. 51 Example A—Discharge oscillations at an applied voltage of 409.5 V. a) As measured, b) Simulation using parameters extracted from A: $V_b' = 389$ V, $V_x = 256$ V, $C = 24.6$ pF, $R_g = 2.2$ MΩ.

All experiments: $R_b = 100$ MΩ, $d \approx 10$ μm. Gas: N_2 at 10 kPa.

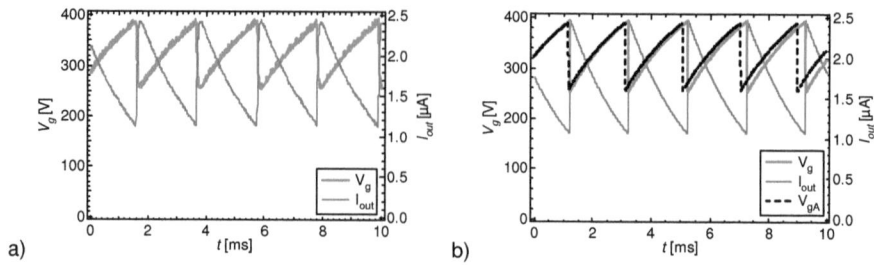

Fig. 52 Example B—Discharge oscillations at V_a = 500 V. a) As measured, b) Simulation using parameters extracted from B: V_b' = 394 V, V_x = 253 V, C = 23.7 pF, R_g = 1.2 MΩ. The dashed line in b) results when the parameters from example A are used and V_a is set to 500 V in the simulation.

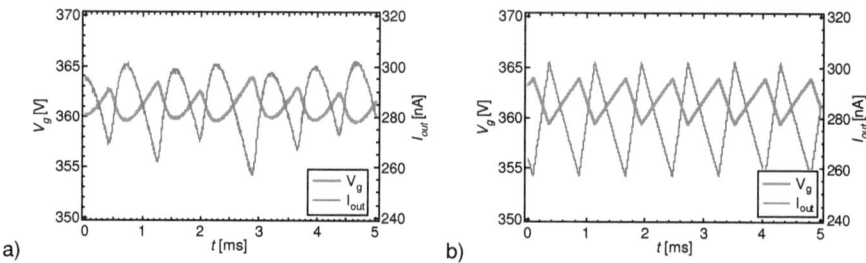

Fig. 53 Example C—Discharge oscillations at V_a = 389.6 V, partly insulated cathode. a) As measured, b) Simulation using parameters extracted from C: V_b' = 363.9 V, V_x = 359.4 V, C = 32.7 pF, R_g = 432 MΩ.

A comparison of Fig. 51 and Fig. 52 shows how the frequency of the oscillation and the circuit current increased with increasing applied voltage, while the level and amplitude of the gap voltage remained almost the same. The gap voltage was calculated from the measured current and the ballast resistance with

$$V_g = V_a - I_{out} \cdot R_b \qquad (5.12)$$

Example C (Fig. 53) shows a measurement where the cathode was partly insulated, thus limiting the spread of the glow. Here, at a lower V_a, the frequency is higher, while amplitude and current are lower than with non-insulated electrodes. In this case, because of the particular shape of the curve, the parameters for the simulation could not be found from an exponential fit like for the other examples. Here we started from the measured discharge period t_b and ratio between rising and falling current, and calculated the time constants using (5.8). Finally R_g was adjusted to achieve the original frequency. The insulation has clearly resulted in an increased gap resistance due to the reduced discharge volume.

5.3 Simulation of glow discharge oscillations

In Fig. 54 two graphs of measurements of frequency over applied voltage are compared. In the range of these measurements the frequency was proportional to the applied voltage and the inverse of the circuit resistance. This behaviour is predicted by our model, see (5.8), where the frequency is proportional to $1/R_b$, and proportional to V_a for $V_a \gg V_b'$.

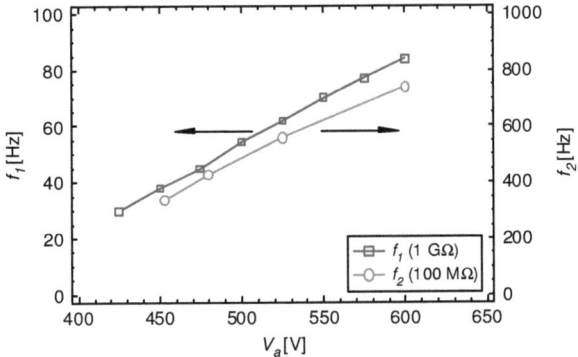

Fig. 54 Measured discharge pulse frequencies over applied voltage.
f_1: $R_b = 100$ MΩ, f_2: $R_b = 1$ GΩ. $p = 10$ kPa of N_2. Electrode type: Plan.

5.3.4 Discussion

Parameter extraction

Table 14 gives an overview of the parameters extracted from the example measurements.

Table 14 Fitting parameters for the measured oscillations of examples A to C in Fig. 51 to Fig. 53.

Example	V_a [V]	V_b' [V]	V_x [V]	τ_{open} [ms]	C [pF]	R_g [MΩ]
A	409.5	389	256	2.46	24.6	2.2
B	500.0	394	253	2.37	23.7	1.2
C	389.6	363.9	359.4	4.91	32.7	432

The extracted parameters were used for calculations of the gap voltage and circuit current curves that are shown next to the measurements in the respective figures. By adjusting the parameters, the simulation can be adjusted to even better match the measurements. The shape can be further improved by introducing inductivity to the model circuit. This does not significantly improve the quantitative accuracy of the model and was therefore omitted.

5 Micro gas discharge devices

The influence of the model parameters

If V_b' and V_x change in the model, the effect is obvious from Fig. 48. Ideally the applied voltage V_a should not influence V_b' and V_x. But in fact the voltage where breakdown is observed will increase with V_a, because breakdown takes some time to develop, so the gap voltage will "overshoot", also due to the circuit inductance. These effects are neglected in the model.

A greater V_a speeds up the processes in the gap and leads to a higher frequency. The influence of V_a on the frequency can be estimated with (5.8). In the calculation, f increases approximately linearly towards infinity for large V_a, while f tends quickly to 0 when V_a goes towards V_b'. However, the model is only valid in a certain range of experimental parameters: For V_a, the lower limit is close to V_b', where the pulses in discharges become irregular before ceasing. The upper limit for V_a where oscillations are present is given by:

$$V_a < V_d \cdot \left(\frac{R_b}{R_g} + 1 \right) \qquad (5.13)$$

At higher V_a, the gap voltage remains above V_x, the switch remains closed, the oscillations stop. According to this model, a steady discharge could therefore be obtained by increasing V_a or decreasing R_b. However, the theoretical limit for V_a under the conditions of experiment A would be 11900 V. Alternatively R_b would have to be decreased to 1.3 MΩ for a steady current. Both conditions can in practice not be fulfilled, because the discharge turned into arcing, destroying the electrodes, before the oscillations could stop. In discharges at larger scale we have observed a mode change to a silent glow at high voltage, see §6.1.1. Similar observations have been reported by Raizer [RAI87] and Akishev [AKI99], for example.

Increasing C or R_b decreases f. A higher capacity corresponds to a higher ionization ratio or a greater ionized volume in the gap. In the same gap, this can mean a greater extension of the cathode glow, for example. Therefore the frequency is inversely proportional to p and d as a first approximation, but changing these parameters has also an effect on all the other parameters of the experiment, in particular on V_b' and V_x. This influences the curve shape but has a lesser impact on the frequency.

A partial insulation of the cathode limits the spread of the glow across the cathode. R_g increases accordingly. C remained almost constant in our model, which can be explained with an increased ion density in the glow. A high ion density could also facilitate breakdown in a partly recovered gap and therefore be the reason for the observed reduction of V_b' in experiment C. V_x, on the other hand, was higher than before. With the non-insulated cathodes in experiments A and B, V_x was very close to the theoretical Paschen minimum of 251 V given by Naidu [LUX01]. A glow can only exist when there is enough space between the electrodes for a cathode layer to form at a given voltage. The thickness of this layer decreases with increasing voltage. Without insulation, the cathode layer has enough space available to spread as necessary when the gap voltage decreases. When the available space is reduced by a partial insulation of the cathode, the cathode layer is only sufficiently thin at a higher voltage. We suggest this as an explanation for the increase of V_x in example C.

That the oscillation amplitude reduces when the cathode is insulated, gives a hint on the oscillation mechanism. Likely the cathode glow spreads to a maximum extension, then the space charge build-up shields the field between cathode and anode so much that the field strength becomes insufficient to sustain the glow. After the space charges have sufficiently reduced, the glow spreads again. A greater glow volume would consequently have a lower oscillation frequency, because it contains more space charge that takes a longer time to build up and to vanish. The observed glow extension depended on the pressure and increased at lower pressure, while the frequency of slow oscillations was also proportional to the pressure. This theory is further supported by the fact that the frequency increased with the current, since a higher current means a faster replenishment of space charges.

There was always a lower amplitude high frequency oscillation present, sometimes superposed over low frequency, high amplitude pulses. The mechanism for those is a different one, probably it is the standard glow discharge oscillation, as it also occurs in corona and dark discharges, well described by Ercilbengoa [ERC01]. It must be kept in mind, that our discharges are not taking place in homogeneous fields. So there will be a corona-like part in our glow discharges.

Conclusions from discharge simulation

Using a simple electric circuit, the influence of a number of parameters on an oscillating glow discharge was modelled. One measurement of a discharge in a given gap is sufficient to determine the values of the principal discharge parameters to be entered into the model. Then these parameters can be varied in the model and their effect studied.

Our model is not concordant with the physical processes in a discharge gap, and not all interdependencies, like the influence of the gap current on the gap resistance and capacity, are covered. Still the resulting curves coincided with measured oscillations. The effects of changing parameters in the model reflect well the effects observed in actual discharges. Our model is therefore a useful tool for the analysis and understanding of discharges as well as for the design of experiments and improvement of discharge devices.

5.4 Conclusions

5.4.1 The influence of experimental parameters

Gas type

The type of gas near the discharge electrodes influences the discharge at several levels. First of all gas molecules adhere to and are potentially even absorbed by the electrode surface. Like this, the work function of the surface changes, changing the electron emission properties of the surface. During discharge or development of the same, the secondary emission, described by Townsend's second coefficient γ, changes, because now not only the surface atoms are hit by electrons and ions, but also the gas molecules on the surface. Even more dramatic effects can be expected, when gas molecules react with the surface material, producing an additional gas or surface component. Such reactions are to be expected, since reactive ions are readily created in a plasma, especially in air.

Apart from surface effects, the gas has a bearing on the volume discharge. Be it by electronegative gases that capture electrons, thus reducing the number of free electrons in the plasma, or more directly by the "ionizability" of the gas, determined most of all by the gases ionization potential. These effects are explained in more detail in chapter 3 of this thesis.

In our experiments we mainly used Ar, N_2, and air. Ar is a noble gas that is relatively easy to ionize, breaks down at a low voltage, and even the ions, Ar^+, are not reactive. N_2 was mainly used because it is the major component of air. N_2 breaks down at a higher voltage than Ar, and the range in which stable glow discharge was observed was smaller. The stability of discharges was generally lower with N_2 than with Ar. Natural air is the most problematic of the gases we used. Since it is a mixture of gases, and also contains significant amounts of water vapour, which is known to affect discharges considerably, its properties and behaviour in a plasma is less easily understood and predictable. We did not evaluate the impact of water vapour on our experiments, but some instabilities might well be due to this factor. Synthetic air, as used in some experiments, contains N_2 and O_2 only, along with traces of other gases and vapours. Here, in a few experiments, we found no obvious difference compared to the behaviour of pure N_2.

Gas pressure p

According to similarity laws and Paschen's law, there is no difference for breakdown if the pressure is changed, or the electrode distance. There is a difference in inhomogeneous fields between the electrodes we used. The breakdown voltage also changes with the inhomogeneity of the field as discussed by Held [HEL97]. Once a discharge has started and a glow has developed, the glow spreads partly across the cathode. The pressure p and the applied voltage V_a determine the extension of the glow. While the glow is confined to a volume between the electrodes and near the gap at a high pressure, the glow spreads more and more over the cathode when the pressure is reduced. With our micro electrodes, the field strength is higher directly on the edge, but then reduces quickly with the distance from the edge, as illustrated in Fig. 31 on page 71. A few μm away from the edge the field strength resulting from the applied voltage is not sufficient to cause breakdown or maintain a glow by itself. The spread of the glow must therefore be due to the additional field between the space charges in the glow and the cathode surface in the cathode fall.

The effect of p on the current during a stable glow discharge is presented in Fig. 55. We believe that the decreasing current towards higher pressure is mainly due to the reduced glow area.

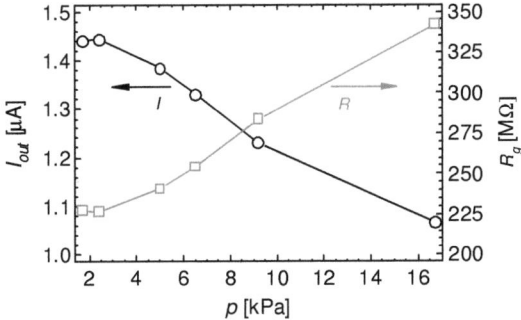

Fig. 55 Discharge current and gap resistance vs. pressure in N_2. The pressure was reduced stepwise during a stable (oscillating) glow. The applied voltage was 470 V. Electrodes: Plan, $d > 1$ µm after initial sparking, $R_b = 100$ MΩ.

For the oscillation frequency of corona discharges, Ercilbengoa [ERC01] observed an increase with pressure (apparently linearly).

Electrode distance d

We have seen that the gap distance must be larger then the Debye length λ_D in order that a glow discharge can be established between the electrodes. Calculations showed that gaps of the order of 10 µm are sufficient if the electron density is sufficiently high ($n_e > 10^{16}$ m^{-3}), see §4.1.1. The formation of a double layer as proposed by Ercilbengoa et al. on the other hand seems unlikely in such small gaps and we have not observed the effects of such a layer.

The Paschen minimum for most gases lies between 0.5 and 1.5 Pa×m. At atmospheric pressure $p = 10^5$ Pa, therefore 5 µm $< d_{min} <$ 15 µm. This is the range of electrode distances for homogeneous fields. Since, as found by Held [HEL97], the position of the minimum V_b is slightly shifted towards greater pd factors, our non-plane-parallel electrodes should have a slightly greater distance. We used distances from 1 to 50 µm first, then concentrated on 3 to 24 µm, still covering the whole range of interesting d. Below this range we enter the domain of vacuum discharge, where the gas is not so important anymore. Table 15 presents extrapolations from one of the first micro glow experiments according to the pd similarity law. This gave a first indication that operation at atmospheric pressure should work for $d = 1$ to 2 µm.

Table 15 Extrapolation of pressure range to achieve stable glow for different electrode distances, taken from a measurement at 10 µm in N_2. pd_{min} = 0.01 Pa×m, pd_{max} = 0.18 Pa×m.

d [µm]	p_{min} [kPa]	p_{max} [kPa]
1	11.0	186.0
2	5.5	93.0
5	2.2	37.2
10	**1.1**	**18.6**
20	0.55	9.3
50	0.22	3.7

Ballast resistance R_b

The ballast resistance R_b firstly limits the current during discharge. Secondly it affects stability. Depending on the V-I characteristics of the system, the discharge goes through stages from dark to arc discharge, compare Fig. 11 on page 38. The lower R_b, the more quickly these stages are passed when increasing the applied voltage, and the sooner the range of abnormal glow is reached, so that sparking becomes more likely. A high R_b on the other hand does not allow for a stable discharge when it is too high compared to the gap resistance during discharge.

Equation (5.13) gives an indication of how R_b should be chosen to achieve a non-oscillating glow. This equation has been confirmed by experiments. Microelectrodes cannot stand the high current resulting from the low ballast resistances necessary for a mode switch. On the other hand, in most cases the oscillations do not interfere with the purpose of the discharge and does not need to be considered an "instability". The best results with discharges in micro gaps were achieved with a ballast resistance of 100 MΩ.

To avoid high capacities between the discharge gap and R_b, the resistance should be close to the gap.

5.4.2 Conclusions from microionizer experiments

The microionizer structures Plan allowed to measure breakdown voltages in small gaps and to perform discharge experiments to find conditions for stable glow with a current of up to several µA. The *pd* range for glow with planar electrodes was approximately from 0.005 to 0.2 Pa×m in gaps from 3 µm to 50 µm in N_2, and from 0.02 to 0.14 Pa×m in Ar. In Ar, discharges were in general much more stable, while the stability was low in air. The glow range depends on the geometry of

5.4 Conclusions

the gap: With the thick Si-Plan electrodes stable glow was achieved at higher pd, up to 4 Pa×m in Ar, 1.6 Pa×m in air. Here the glow stability also depended on electrode activation. The high amplitude oscillations encountered with glow in Plan structures reduced significantly, when we partly insulated the cathodes. This success inspired the design of planar electrodes insulated with a layer of SiO_2 for I-Plan. The lifetime of I-Plan electrodes were extremely short due to destruction of the thin Au layer and formation of a short circuit by deposited metal. The solution was found when we fabricated 50 µm thick electrodes for Si-Bulk. These electrodes, suspended far above the substrate, survived discharge experiments for over 14 hours in laboratory air without severe damage. On the electrodes of Si-Bulk we observed the growth of whiskers and other deposits, and the formation of pores in the cathode. The growth of whiskers leads to an activation of the gap, which leads to a reduced breakdown voltage and, under certain circumstances, to a stabilisation of a glow. After a long operation time, whiskers might cross the gap and lead to a short circuit. A disadvantage of the Si-Bulk design is the difficulty of insulating the electrodes away from the gap to keep the glow from spreading and to avoid sparking to distant spots. Even if spreading is not "dangerous" and remote sparking is not such a threat to lifetime for Si-Bulk as it was for the other designs, the aim for the electrodes to be used as point sources for ions is compromised.

When the ionization only serves as a source for ions that are analysed subsequently by other means, then the absolute stability of the discharge is of minor importance, especially in pulsed operation. It would only affect the accuracy of a quantitative measurement. Such a quantitative inaccuracy can be introduced by:

- Variations and noise in the source voltage
- Movements, vibrations of the electrodes
- Unstable gas flow
- Temperature change
- Change of matrix gas composition, especially humidity
- Electrode contamination and degradation

Electrode degradation, including long term contamination, is a slow process compared to the duration of a measurement, and can therefore be neglected if a quick calibration just before the analysis is done. Electrode contamination can be a short term effect, when molecules adsorb and desorb quickly to and from the electrode surfaces thus influencing the discharge.

6 Ion extraction, filtering and detection

A few general points about mass spectrometry and ion mobility spectrometry were mentioned in the introduction to this thesis. In this chapter we present our own developments towards a miniaturised ion mobility spectrometer. The components we developed and used are introduced and measurements presented.

6.1 Miniature ionizer

6.1.1 Glow discharge in a miniature ionizer

For our model ion mobility spectrometer we used an ionizer made from two pins of a standard connector array that were bent towards each other to create a discharge gap of approximately 50-100 µm. We experimentally determined the circuit parameters for a stable glow discharge of these miniature electrodes using the electrical set-up sketched in Fig. 56.

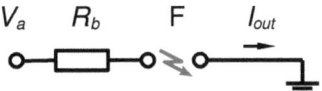

Fig. 56 Circuit for stability experiments with a miniature ionizer. V_a: applied voltage, R_b: ballst resistor, F: discharge gap, I_{out}: output current.

Fig. 57 shows how the discharge current and oscillations depended on the ballast resistance.

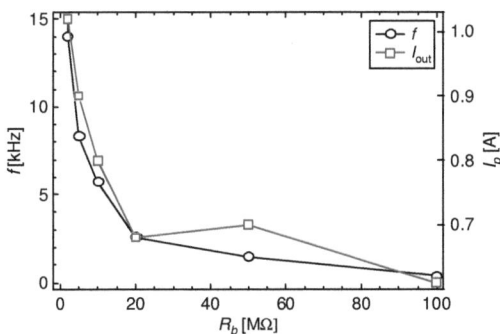

Fig. 57 Graph of frequency and peak current vs. ballast resistance. At $R_b = 1$ MΩ, the oscillations cease.

6 Ion extraction, filtering and detection

With decreasing resistance, the frequency increased and so did the peaks of the output current. At $R_b = 1$ MΩ, the signal switched to a stable mode: there were no more current peaks in the ampere range. The parameters found for a stable discharge with this miniature ionizer are similar to the values found for microdischarges as presented in §5.3.4. Furthermore, the observation of a mode switch to stable glow at $R_b \approx R_g \approx 1$ MΩ confirms the conclusions from the model presented in §5.3. There, a mode switch was predicted at $R_b = 1.3$ MΩ ($V_a = 409.5$ V, $d \approx 10$ μm). The current measured as a function of the voltage drop over the electrode gap V_g during stable glow is shown in Fig. 58. V_g was calculated using $V_g = V_a - IR_b$.

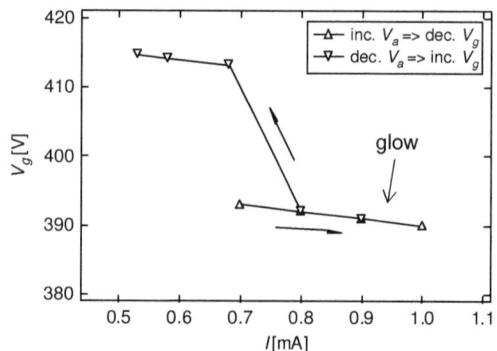

Fig. 58 Graph of the gap voltage V_g against the circuit current I. The arrows indicate the direction of the voltage increase or decrease. $R_b = 1$ MΩ.

Comparing Fig. 58 with the theoretical V-I characteristic in Fig. 11 on page 38, confirms that we are measuring a stable glow, as the gap voltage is quite stable in the shown current range. A sparking region was always passed when increasing the voltage to produce a stable glow.

6.2 Ion extraction

6.2.1 Ion extraction with and without grid

To filter and detect ions created during an electrical discharge, they must first be extracted from the discharge and drawn towards the detector. In order to separate ions, a pulse of ions must be created. We examined two modes of operation:

1. Applying a pulsed field during a steady glow discharge
2. Applying a steady field during a discharge pulse

The first approach, pulsing an extraction field with a grid, has the advantage that the glow discharge remains undisturbed. The second approach, pulsing the discharge, has the advantage that no grid is necessary that obstructs ions on their way into the separator. Another advantage of pulsing the discharge is, that the potentially destructive discharge is only burning for very short times, which can also save power in a portable device. On the other hand, even more destructive initial sparks might appear during the ignition of the discharge. Also, a much higher voltage must be applied to ensure a rapid ignition. Both methods were applied in our experiments.

6.2.2 Grid

Standard grids are fine metal gratings, but grids can assume various shapes. One shape we used was a copper sheet with single holes of diameters from about 1 to 5 mm. An even simpler grid was made from a single wire mounted near the discharge slightly below the central axis of the spectrometer. Like this, the ions were not obstructed, but deviated depending on the potential applied to the wire.

6.2.3 Pulsing

Pulse application

There are several possibilities of how and where a pulse can be applied:

- Applied to a grid:
  The pulse switches the grid potential from a level that holds up ions to one that allows ions to pass
- Applied to discharge electrodes:
  a) Both electrodes 0, the pulse sets one electrode on a high potential causing discharge
  b) Both electrodes high, the pulse sets one electrode to 0 causing discharge
  ("0" = common ground of the setup)

For ionization, a high voltage is necessary, at least 200 to 300 V, depending on the gas. To create a constant voltage of 300 V or more is possible with a simple circuit, but drains a standard battery quickly. The battery lifetime can be improved by using only pulses of a high voltage for ionization. For this reason pulsing the discharge is preferable over pulsing a grid near a continuous discharge.

The way pulses were applied is illustrated in the description of the experiments below.

Pulse generation

We fabricated two different pulse generators of which the circuits are presented in the appendix. One generated a high voltage pulse by means of a differentiator circuit. The square pulse of an oscillator was differentiated and the resulting low voltage peaks were transformed by 12:220. This circuit was able to give millisecond pulses of about 1000 V.

6 Ion extraction, filtering and detection

In the second circuit pulses from a function generator controlled a MOSFET transistor. The transistor switched voltages up to 1500 V for a pulse duration down to 2 µs.

Pulsed discharges

When pulsing the discharge instead of a grid to create pulses of ions, the ignition and short term stability of the discharge becomes an issue. Fig. 59 shows two examples of discharge pulses. It shows that by increasing the applied overvoltage the stability is improved, i.e. the variation of the number of ions in the gap during the discharge as well as the variation of the total number of ions created during a pulse is reduced. The shorter the applied pulse, the more the improvement will be pronounced. In our experiments the pulses were well repeatable.

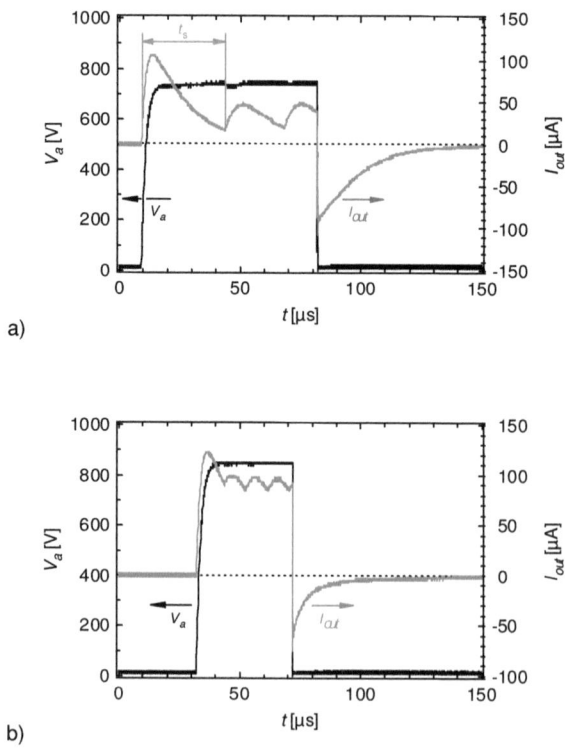

Fig. 59 Pulsed discharges with miniature electrodes at different voltages. $R_b = 2.29$ MΩ. a) $V_a = 740$ V. The discharge starts with the pulse but oscillates with a high amplitude and low frequency. b) $V_a = 860$ V. At a higher pulse voltage the oscillation amplitude decreases, the frequency increases.

6.2 Ion extraction

When changing the pulse length, the time until the discharge had stabilised (t_s) stayed the same. t_s is the time until the discharge has reached steady state (steady oscillations). In this case, it is simply the time until the second oscillation pulse. Increasing the pulse voltage decreased t_s. t_s was 34.6 µs at 740 V in, 11.9 µs at 860 V in Fig. 59 b). For other pulses we measured 12.5 µs at 850 V, 6.5 µs at 1000 V.

6.2.4 Detector and signal amplification

For the detection of ions we used metal plates of various materials and diameters. At first, a gold plated connector of a BNC socket with about 1.5 mm diameter was used. The diameter was then increased with additional plates made of brass or copper. An increased detector surface has the advantage of more ions being captured thus increasing the sensitivity, but shielding became more difficult.

The signal was amplified with a pA amplifier/electrometer connected to an oscilloscope. With a self made amplifier currents from 0.1 pA to 2 µA could be measured. A commercial amplifier was able to handle a greater current range, but proved less reliable: the offset was not stable and the amplifier was too easily damaged.

6.2.5 First extraction experiment

We tested extraction of ions from a micro glow discharge, and switching the ion flow using a coarse grid as a shutter as shown in Fig. 60. The fore mentioned miniature ionizer was used in this experiment. We now applied two resistors R_{b1} and R_{b2} of value 910 kΩ to stabilise the glow discharge and to keep the average plasma potential at half the applied voltage. A stable DC glow discharge (< 0.5 % oscillations) was obtained in laboratory air at about 800 V and 0.27 mA DC. The gap resistance during discharge was therefore about 1.1 MΩ.

A pair of connector pins, 2.54 mm from the discharge gap, were extended and served as a grid. By applying an electric field between the glow region and the detector, ions were extracted through the grid. Fig. 60 shows the electrical scheme of the setup.

6 Ion extraction, filtering and detection

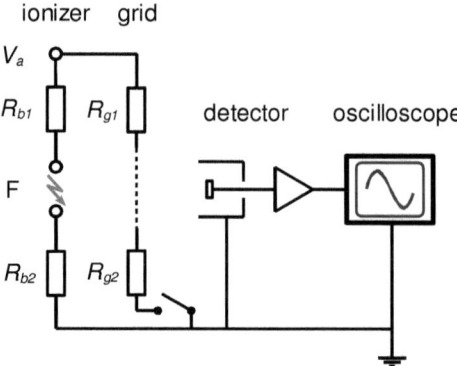

Fig. 60 Electrical scheme of ion extraction set-up. The switch is hand-operated.
$R_{b1} = R_{b2} = 910$ kΩ, $R_{g1} = 600$ kΩ, $R_{g2} = 300$ kΩ.

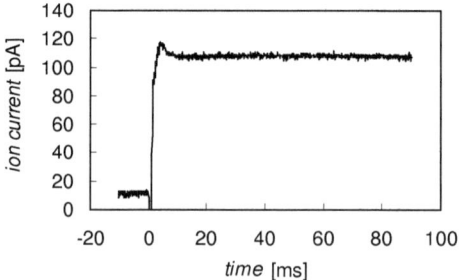

Fig. 61 Net ion current when opening the electrical grid shutter at $t = 0$ s.
$d \approx 100$ μm, $V_a = 800$ V, $V_{grid} = 267$ V.

The detector electrode had a diameter of about 1.5 mm and was positioned 10 mm from the grid. With a push button the grid was switched between the applied voltage and one third of it. The amplified current was displayed with an oscilloscope. Switching the grid potential to a high potential stopped the detector current almost completely. This switching process is illustrated in Fig. 62. The net detected ion current is shown in Fig. 61 ($I_{off} \approx 10$ pA, $I_{on} \approx 110$ pA). By blocking the air path with an insulator we verified that the detected current was indeed ion flow. The net ion current is defined as the measured current through a free path minus the current measured when the path is blocked. We thus subtract a large part of switching noise from the signal.

Comparing the ionizer current of 0.27 mA and the detected current of 110 pA we achieved an *ion yield* of 4×10^{-7} in this case. We define the ion yield as the detected ion current divided by the current through the discharge. This number gives an indication of the overall efficiency of the system, including the efficiency of ion creation in the discharge, of extraction of ions from the discharge, the losses in the drift chamber, and the detector efficiency.

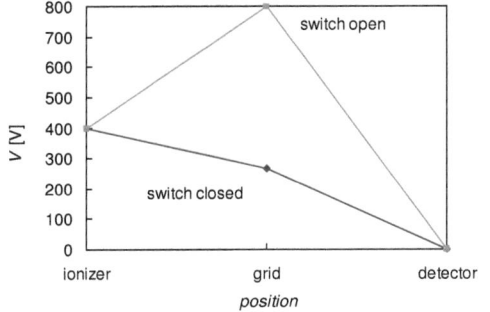

Fig. 62 Illustration of grid operation: When the switch is closed, the grid potential is reduced to a value defined by R_{g1} and R_{g2}.

Here: $V_a = 800$ V, $R_{g1} = 600$ kΩ, $R_{g2} = 300$ kΩ. The grid potential with closed switch allows ions to pass.

6.3 Miniature ion mobility spectrometer (IMS)

We constructed a simple miniature ion mobility spectrometer (IMS) with a planar filter structure. A planar structure is easy to fabricate and best adapted to micromachining. It was possible to extract ions to a detection electrode with a certain delay as they go through an ion filter. A pulse generating circuit was designed using simple, readily available components, to provide pulses for the extraction of ions. The final set-up was enclosed in a box to minimize effects due to air currents and also to test various gases. This setup served to verify the feasibility of the proposed micro analyser. In this system it is much easier to start, maintain and observe the discharge, a greater amount of ions can be extracted more easily, and experiments with the ion filter and detector are done with less effort and more quickly than with a microsystem. A picture of a detail of the system is shown in Fig. 63, the electrical scheme is given in Fig. 64.

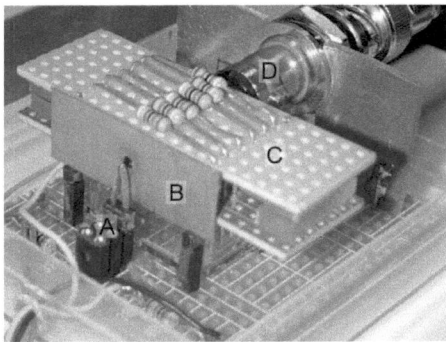

Fig. 63 The experimental set-up for a miniature spectrometric gas analyser shown without box cover. A: ionizer, B: grid, C: drift chamber, D: detector.

6 Ion extraction, filtering and detection

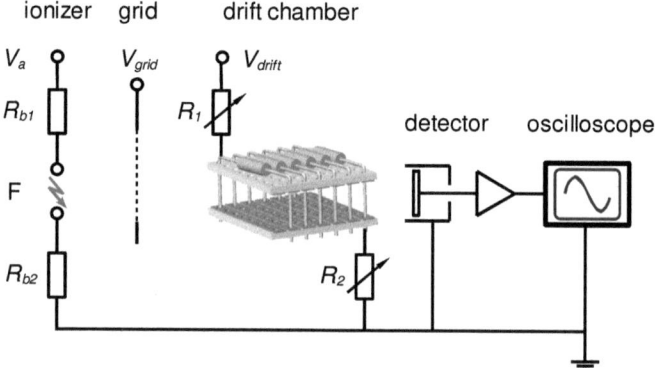

Fig. 64 Illustration of the experimental set-up for a miniature spectrometric gas analyser. The grid is switched by a pulse generator. Alternatively the discharged is pulsed; then no grid is used. The variable resistor R_v regulates the potential at the exit of the drift chamber.

The same metal pins and resistors as before were used as the ionizer. A metal plate with a Ø3 mm hole served as a grid as shown in Fig. 63. The distance between ionizer and grid was about 2 mm. The resistors used for the drift chamber were 6× 910 kΩ. Ions were extracted through the grid by applying an electric field between the glow region and the detector. The detector electrode had a diameter of about 7 mm. The complete distance from the ionizer to the detector was about 3 cm. Fig. 65 and Fig. 66 illustrate the operation when pulsing the grid and when pulsing the ionizer.

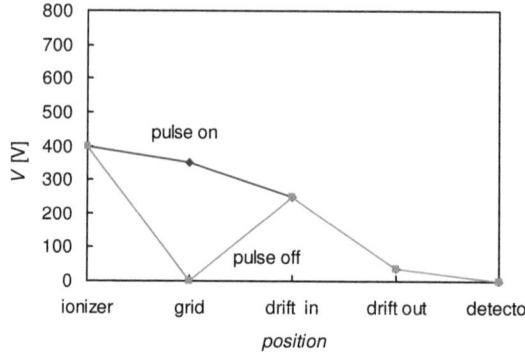

Fig. 65 Illustration of pulsed grid operation: When the pulse is off, the grid potential is 0 V, the ions are caught by the grid. When the pulse is on, ions may pass.

$V_a = 800$ V, $V_d = 250$ V, $R_{b1} = R_{b1}$
$= 910$ kΩ, 600 kΩ, $R_v = 910$ kΩ.

6.3 Miniature ion mobility spectrometer (IMS)

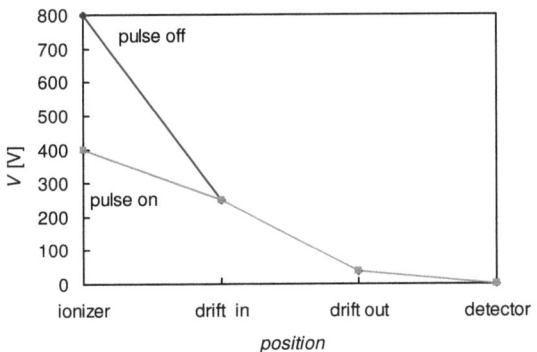

Fig. 66 Illustration of pulsed ionizer operation: When the pulse is on, one of the ionizer electrodes is set to 0 V, a discharge develops and ions are created. When the pulse is off, both electrodes are on a high potential, no more ions are created, while the existing ions accelerate towards the drift chamber.

$V_a = 800$ V, $V_d = 250$ V, $R_{b1} = R_{b1}$ = 910 kΩ, 600 kΩ, R_v = 910 kΩ.

6.3.1 The drift chamber

The ion separation due to different mobilities takes place in the drift chamber, see Fig. 64. This chamber consists of two printed circuit boards (PCB) with parallel metal strips. The boards are mounted on top of each other with a distance of 6 mm to form a drift chamber. The chamber is 1.8 cm long and 2.5 cm wide. Each strip of the upper PCB is soldered to the corresponding strip of the lower PCB to provide a symmetric drop across both the sections. A nearly uniform electric field thus develops across the drift chamber to provide uniform drift velocity. The field can be considered uniform as an approximation as the simulation of the field shows, see Fig. 67 and Fig. 68. In existing ion mobility spectrometers either a series of rings similar as in Fig. 67 b), or a tube coated with a resistive layer is applied to achieve a homogeneous field inside the drift chamber. If simply two parallel plates were used on the ends of the chamber, the field would be disturbed by the ions. The drift field is even more disturbed by the other surrounding walls of the chamber that also carry a potential, if the length of the chamber is not small compared to its lateral dimensions.

6 Ion extraction, filtering and detection

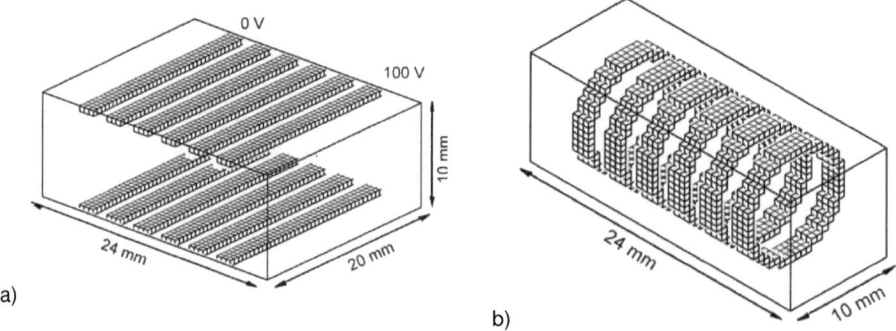

Fig. 67 a) Geometry: 2 mm wide strips, 1 mm distance, 10 mm high.
b) Geometry: 2 mm wide strips, 1 mm distance, 10 mm diameter. This corresponds to the usual design for IMS drift chambers [EIC01b].

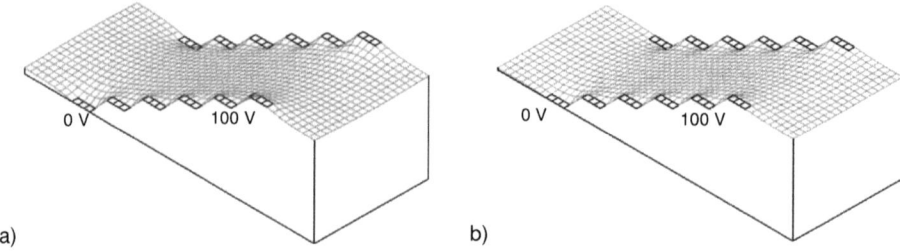

Fig. 68 a) 3D illustration of potential energy (voltage) at and between the strip electrodes.
b) The potential energy in ring shaped electrodes. (Almost the same as a)!)

6.3.2 Measurements with the miniature analyser

All measurements were done at room temperature and at atmospheric pressure.

In the first experiments, the pulse generator generated a pulse going from 900 V down to 500 V to let in a pulse of ions for a period of 2 ms. A measurable delay was observed for ions of N_2, air and Ar. A double peak was observed for Air and N_2 suggesting two different ion species, see Fig. 69. The double peak could be due to two different ionized species of the same gas or also due to two species of different gases. Since the two peaks did not appear in all measurements, it is possible that this observation was an artefact.

6.3 Miniature ion mobility spectrometer (IMS)

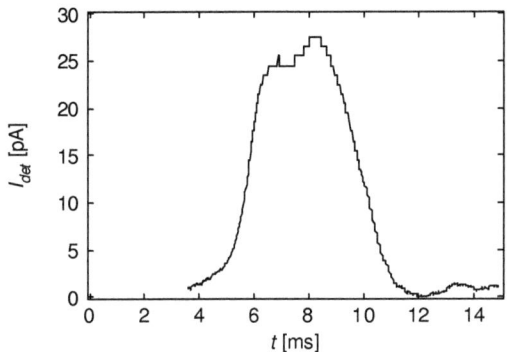

Fig. 69 Overlapping peaks detected in laboratory air. During the first milliseconds, the signal is still strongly disturbed by the pulse applied to the grid.

$V_a = 900$ V, $V_d = 500$ V, $t_g = 2$ ms.

To compare the measured pulse with the theory described in §2.2.3 on page 16, we apply the parameters $l_d = 3$ cm, $V_d = 500$ V, $t_g = 2$ ms in equation (2.39). The drift time is then $t_d = 6.4$ ms for a mobility of 2.8×10^4 m^2/Vs, and the pulse width at the detector $w_{1/2} = 2.01$ ms. The resolution R is found from (2.42) to be 3. Theoretically, the highest resolution of $R = 13$ would have been achieved with $V_d = 70$ V. t_g makes up the major part of the calculated pulse width $w_{1/2}$ and was obviously too long to achieve a good resolution. The theoretical pulse width of 2 ms suggests that the signal in Fig. 69 shows indeed two separate peaks. The second peak would originate from ions with a mobility of about 2.2×10^4 m^2/Vs.

Influence of the experimental parameters

Drift potential V_d

Peak width and delay times depend on the drift potential V_d. For higher drift potentials, the observed delay time was lower and the peak was higher, see Fig. 70. The highest peak obtained was of the order of tens of pA.

6 Ion extraction, filtering and detection

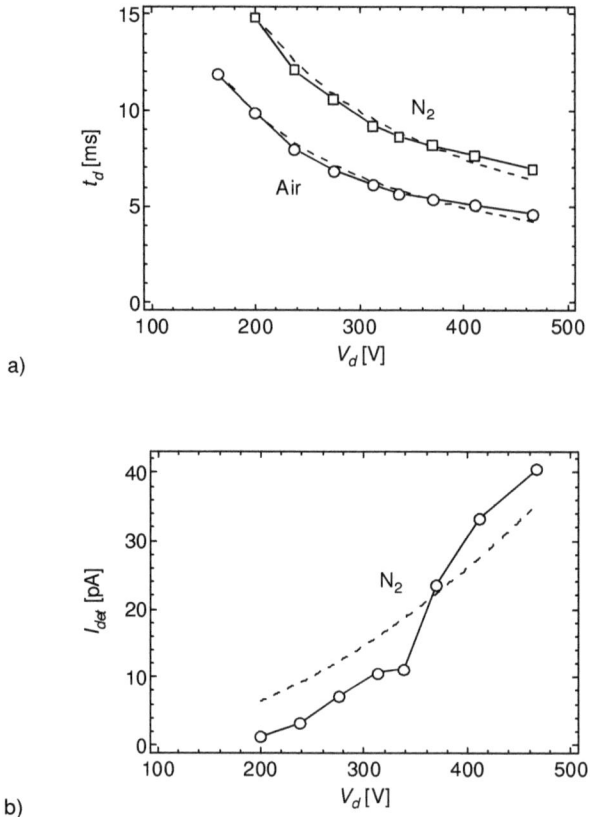

Fig. 70 a) Drift time over drift voltage in air and N_2. A voltage of 90 V was added to the voltage applied to the drift chamber to compensate for drift region variations (three drift regions: Ionizer – drift chamber / chamber / chamber – detector). The dashed lines are theoretical curves (eq. (2.35) with $\mu_{air} = 2.0\times10^{-4}$ m^2/Vs, $\mu_{N2} = 1.3\times10^{-4}$ m^2/Vs.
b) Detector peak current over drift voltage for the measurements in N_2. The dashed curve is a quadratic fit: $a\times V_d^2$, where $a = 1.6\times10^{-4}$.

Drift length l_d

According to equation (2.35) the drift time is proportional to the square of l_d. A measurement of drift time over length is shown in Fig. 71 that is consistent with this theory.

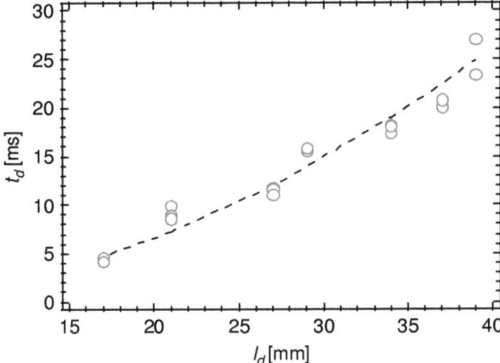

Fig. 71 Variation of drift time with drift length measured in air. The point of 50% of the peak height was used to determine the delay. A grid was used that was switched, not pulsed. $V_a = 1008$ V, $V_d = 321$ V. The dashed line is calculated with equation (2.35), $\mu = 1.9 \times 10^{-4}$ m^2/Vs.

Fig. 72 shows another measurement of t_d over l_d. The mobility found in this experiment was much lower than in the experiment shown in Fig. 70. Mobility and results of mobility measurements are very sensitive to experimental parameters. This makes it difficult to find reliable mobility data in the literature that is applicable to the spectrometer used.

According to an observation by Tabrizchi et al. [TAB99] the total ion current on the Faraday cup depends quadratic on the drift field, which corresponds well to our measurements.

6 Ion extraction, filtering and detection

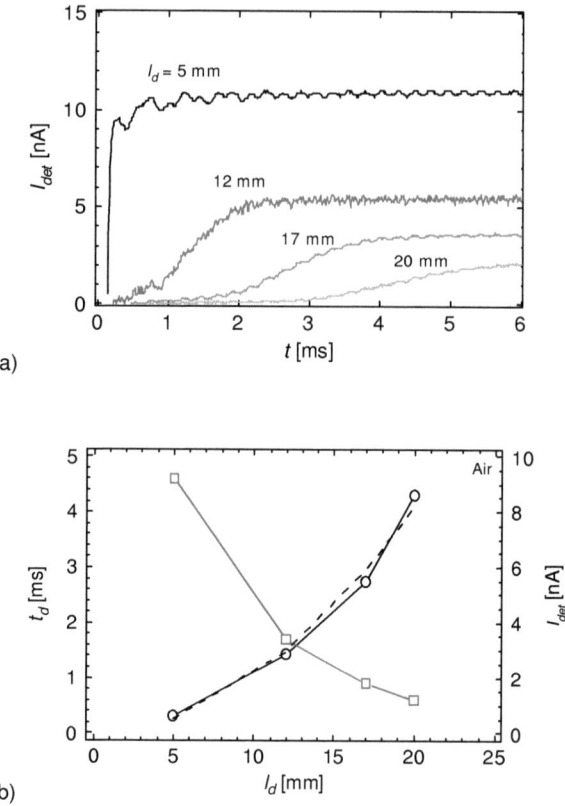

Fig. 72 a) Development of the detected signal as a function of the drift length. $l_d = 5 .. 20$ mm, $V_a = 1300$ V, $V_d = 650$ V. Noise was subtracted from the measured curves. b) Drift time and maximum current as extracted from a). The dashed line is calculated with equation (2.35), $\mu = 1.5 \times 10^{-4}$ m^2/Vs. A switched grid was used, t_d was determined at the point of 50% of the peak voltage.

Ionization pulse length t_{iz}

The influence of the pulse duration t_{iz} on the detected current I_{det} was measured varying t_{iz} from 0.16 to 0.45 ms, see Fig. 73. When decreasing the pulse time, the detected pulse was lower. That the detected current was less is explained by the lower number of ions created and extracted from the ionizer. The drift field is reduced during application of the pulse.

6.3 Miniature ion mobility spectrometer (IMS)

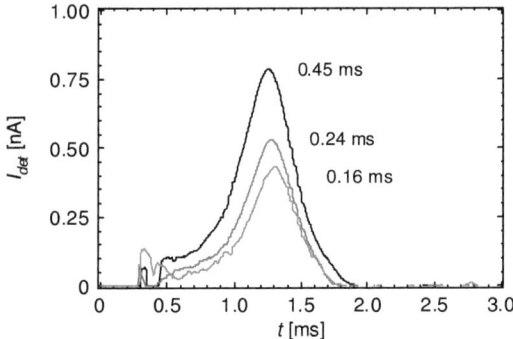

Fig. 73 Detector current over time in air. Both ionizer electrodes were set to $V_a = V_d = 1550$ V, one was pulsed to 0 V for the durations indicated on the graph. No grid was used, noise was subtracted from the measured curves. $l_d = 2$ cm. $R_{0.45\,ms} = 2.7$; $R_{0.24\,ms} = 3.0$; $R_{0.16\,ms} = 3.2$

Table 16 Parameters extracted from the measurement in Fig. 73.

t_{iz} [ms]	I_{peak} [nA]	Area [nC]	t_d [ms]	$w_{1/2}$ [ms]	R	$w_{1/2}/t_{iz}$	$\mu \times 10^4$ [m²/Vs]
0.16	0.45	40	1.31	0.39	3.4	2.44	2.1
0.24	0.53	53	1.28	0.43	3.0	1.79	2.0
0.45	0.79	88	1.25	0.47	2.7	1.04	2.0

Table 16 contains the parameters extracted from the curves in Fig. 73. The best resolution was achieved with the shortest pulse. What is surprising is the small broadening of the longest pulse compared to the others. This suggests that the actual ion pulse width did not depend much on the time of application of the discharge pulse.

Ionized gases

We examined the effect of applying different gases in the drift chamber. Fig. 70 a) already shows a noticeable measured difference between the mobilities of N_2 and air. In Fig. 74 we present two peaks of air and Ar with air as the drift gas, measured four times each to check reproducibility.

6 Ion extraction, filtering and detection

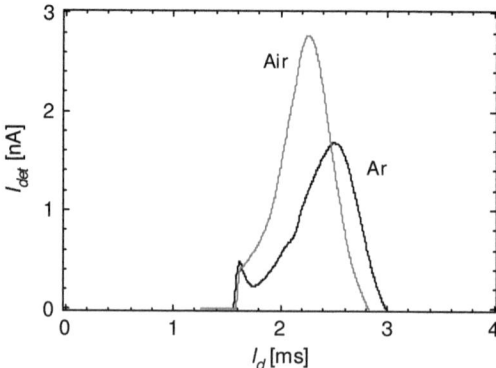

Fig. 74 Signals detected in air as matrix gas with miniature electrodes. To measure Ar, a stream of Ar was directed at the ionizer. $V_a = V_d = 1490$ V, one electrode was pulsed to 0 V, no grid was used, noise was subtracted from the measured curves. $l_d = 2$ cm, $t_{iz} = 0.16$ ms. Extracted parameters air/Ar: I_{peak}: 2.8/1.7 nA, t_d: 2.25/2.49 ms, μ: 1.2/1.1×10^{-4} m^2/Vs.

The two gases can easily be distinguished from their peak height and drift time. However, the resolution achieved here is still too low to distinguish ion species in a mixture.

6.3.3 Microionizers as ion sources

The microelectrodes of type Si-Bulk have a lifetime in air in excess of several hours. It was therefore possible to operate such ionizers in the model spectrometer. Fig. 75 shows a Si-Bulk chip mounted in the IMS. An ion current from a micro discharge pulse is graphed in Fig. 76. In this example no ballast resistance was used, which resulted in an arc discharge. The repeatability and spacial confinement of these discharges were not satisfactory, and the peak resolution was low. But the peak was high, on the order of 10 nA, compared to the current that is usually extracted from radioactive sources, for example, which is as low as 10 pA. It should therefore be possible to find a discharge regime where the pulse is at a lower—still useful—current level, but better controlled, using a suitable ballast resistor and a conditioned micro ionizer.

6.3 Miniature ion mobility spectrometer (IMS)

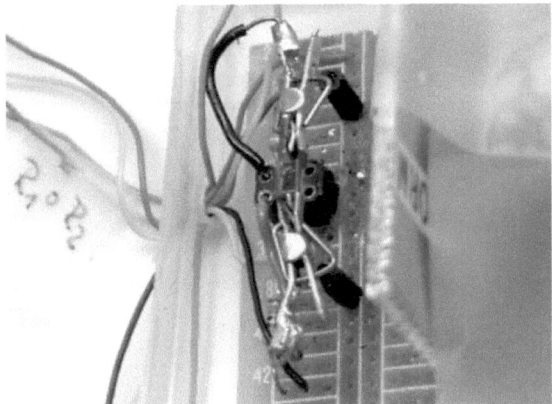

Fig. 75 Microionizer mounted in the miniature spectrometer. The ionizer, type Si-Bulk, is held and contacted by two miniature crocodile clamps.

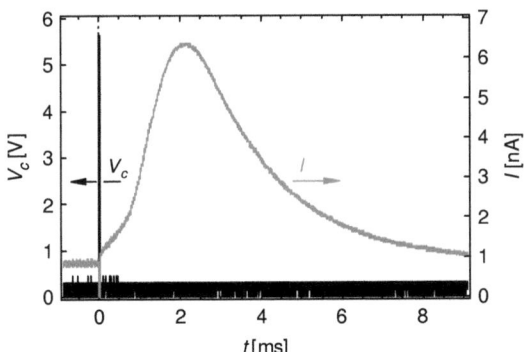

Fig. 76 Ion pulse in air with Si-Bulk electrodes. V_c is the control voltage of the transistor that switches the high voltage. $V_a = 700$ V, $t_{iz} = 20$ µs, $d \approx 3$ µm, $l_d = 1.2$ cm. No ballast resistance added, no grid used.

Optical spectroscopy of a micro discharge

In cooperation with the Plasma Physics Research Centre (CRPP) at the EPFL we measured the optical spectrum from a micro discharge using Si-Bulk electrodes. The discharge was maintained in air at about 750 V. An optical fibre served to feed the light from the discharge to a spectrometer. The resulting spectrum is presented in Fig. 77. The peaks are practically all emitted by neutral N_2 molecules, but also two lines from N_2^+ ions can be seen. There is no trace from O_2, O, N, O_2^+, O^+,

6 Ion extraction, filtering and detection

N$^+$, etc. that we had expected. That they cannot be seen here does not mean that these species did not exist in the plasma. It only means that they were not excited and did not emit light.

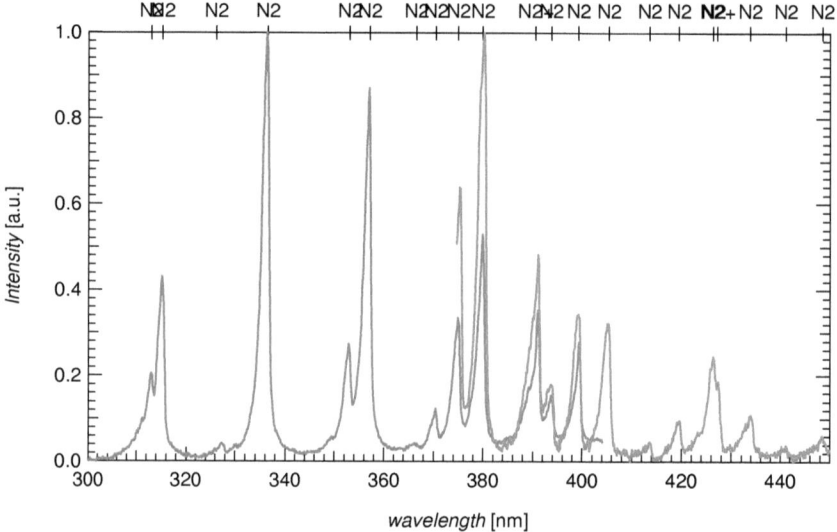

Fig. 77 Optical spectrum from a micro discharge with Si-Bulk electrodes in air. The spectrum was measured in wavelength bands of which two are shown here, partly overlaid in the centre. $V_a = 750$ V, $R_b = 100$ MΩ.

6.4 Conclusions and suggestions

In this chapter we presented our work on a miniature ion mobility spectrometry setup. A stable mode of operation was found for a miniature discharge ionizer, and ions were extracted from the same. This ionizer was then integrated with a simple planar drift chamber and a detector. We were able to detect ion pulses for a pulsed grid and a steady discharge as well as for pulsed discharges. The experiment with this setup allowed us to examine the influence of various parameters on the detected ion pulses. A good agreement with theory was found for example for the relationship between drift time and drift length. What is more, is that we were able to differentiate between different gases, although the resolution of our setup was not high enough to clearly analyse a mixture of gases. The detected ion current was fairly high for miniature as well as for micro ionizers, the signal to noise ratio has therefore not been a problem. The major issue to improve remains the resolution. We identified the following possibilities for improvements:

- improved drift field homogeneity
- reduction of the initial pulse length
- reduction of the pulse extension at the ionizer

6.4 Conclusions and suggestions

- optimization of drift field strength
- control of experimental conditions (composition of drift gas, gas flow, temperature)
- additional "aperture" grid before the detector
- use of advanced operation options used in standard IMS:
  choice of drift gas, unidirectional flow design, Fourier transformation for measurement analysis, agents for chemical ionization

The resolution could be enhanced with a better homogeneity of the drift field. Currently the ionizer is part of the drift field, and even when both electrodes are set to the same potential after a discharge pulse, the electrode tips cause an inhomogeneity of the field near the tips, where most of the ions are located just after the pulse. Their initial path is therefore strongly disturbed. A solution would be to use a pulsed grid, which means to lose some signal strength, or to apply a planar ionizer.

Reducing the pulse length is only useful if the applied pulse length is in the order of magnitude of the detected pulse width. The pulse widths we measured were in the order of milliseconds, while we applied pulses of less than 0.5 ms. Even with much shorter pulses (20 µs) the resolution did not increase significantly. A greater gain can be hoped for with a reduction of the pulse extension, which was nearly 1 mm with the miniature electrodes, and not much less with the microionizers, because these were operated in spark mode, where the sparks went beyond the gap. When the discharge is better controlled and confined to the microgap, a glow would not extend beyond the thickness of the electrodes, which was 50 µm.

The thickness, or "depth" of the ionizers cause an additional broadening: the ions that are near the edge towards the drift chamber see a stronger field than the ones further away, as the illustration in Fig. 78 demonstrates. In the ionizer, not far from the edge, the gap is shielded by the electrodes. Ions inside the gap will not see the field between electrodes and grid or detector plate. To achieve a sharp initial ion pulse, the electrodes should therefore not be much thicker than the gap width.

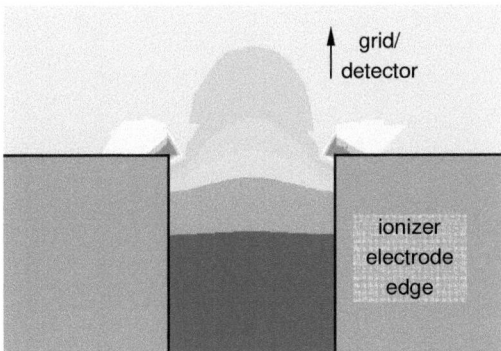

Fig. 78 Electric field in and near an ionizer gap between electrodes at the same potential and a grid/detector plate at a different potential. The coloured/shaded areas show the field strength. It is strong at the corners, becoming homogeneous towards the grid/detector, and low between the electrodes.

6 Ion extraction, filtering and detection

One parameter in the calculation of resolution is the drift field strength, see §2.2.3 on page 14. The resolution depends on the field strength in a complex way and is best found experimentally.

A more efficient ionization could be achieved with the sample gas flowing through a discharge. This would be possible with the ionizers tested so far with an adapted gas flow, and supposedly even more with an adapted gas flow and ionizer.

"Aperture" grids are commonly used in ion mobility spectrometers. Such grids, applied directly in front of the detector, serve to avoid that approaching ions induce a current in the detector from a distance, thus causing a broadening of the detected peak.

7 Conclusions and outlook

Discharges in small gaps of micrometre scale have long been neglected in science and only recently gained broader interest. It was therefore the right moment for us to enter this fascinating field of research to which we hope to have presented a worthwhile contribution with this work.

7.1 What has been achieved

We consider our most notable achievement the fabrication of a micro ionizer that has proven to be stable for more than 14 hours in N_2. Also in laboratory air the ionizer has been operated for several hours. Furthermore, it has been successfully applied as the ionizer in a miniature ion mobility spectrometer. Our contribution to the science of micro plasmas are the observations of the same under the micro- and oscilloscope that we describe in this thesis. We measured breakdown voltages in gaps from 1 to 50 µm at a broad range of pressures in Ar and N_2. Conditions for stable glow in small gaps were found and we can now give approximate ranges for the factor pressure times gap distance pd where glow can be expected in planar micro gaps: From 0.005 to 0.2 Pa×m in N_2, and from 0.02 to 0.14 Pa×m in Ar. With thick electrodes glow was achieved up to greater factors: 4 Pa×m in Ar. In our miniature electrodes pd was about 10 Pa×m. The stability of the glow, as we have shown, depends on more factors than pd. Other factors we examined were the gas type, the gap geometry, and the electrical circuit in which the ionizer is embedded, in particular the circuit resistance. Whether high frequency oscillations in a glow discharge are considered an instability, depends on the application and is otherwise open for discussion. However, we examined such oscillations in some depth and developed a model electrical circuit for their simulation. The model gave a good match with measured oscillations and helped to explain the limits of appearance of oscillations. These limits again were confirmed by experiment.

Our final design of a micro ionizer proved not only interesting in terms of the new experiments that were possible thanks to its ruggedness, but that could also, for the first time, be used as the ionizer in the miniature ion mobility spectrometer (IMS) we set up. For this spectrometer we fabricated and assembled miniature ionizers, grids, drift chambers, detectors, and pulse generators. Eventually we demonstrated the feasibility of the aim of our project by showing gas selectivity in our miniature IMS. The achieved repeatability and pulse resolution were not yet satisfactory. We therefore worked out a number of suggestions that should help to make our miniature spectrometer a useful demonstrator on the way towards a new integrated micro gas detector.

For scientists who wish to enter the area of micro plasma research, the basic knowledge we covered in the first chapters of our thesis will hopefully be a helpful support. In addition, our extensive collection of literature references may provide a basis for an efficient entry to the subject

7.2 Outlook

Our accomplishments present a contribution to micro plasma research and a first step towards a novel low cost portable gas detector. On both levels there are ample opportunities for continuation. For research we have shown that microdevices can be an appropriate base for the examination of micro plasma. There remain a number of parameters which we have not or only marginally regarded in our work, but that have a strong influence on discharges: temperature, humidity, and the materials and surface state (adsorbates) of the electrodes.

The miniature IMS experiments have shown that it makes sense to move on and integrate our micro ionizer, or an advanced version of it, into a miniature gas detection system. To integrate a high resolution IMS gas analyser—this is one conclusion from our experiments and reasoning—is an aim that is too ambitious to pursue. For our devices, suggestions for short term improvements can be found in the conclusions of the chapters about our experimental work. For the work that is further away, we suggest the following:

1. To involve experts of ion mobility spectrometry in the project. More than two decades of ion mobility and more than half a century in mass spectrometry research has accumulated so much knowledge, that an expert can help to avoid important design mistakes at this early stage and significantly reduce the time it will still take until the fabrication of a detector demonstrator.

2. To examine alternatives for the discharge type. Electrical discharges have proven, not only in our work, to be suitable for ionization in IMS. For this thesis we had to focus on one discharge type, the DC glow discharge. We defend that this was a good decision under the circumstances. We could not prove that DC glow is the best choice for the final IMS device, and therefore suggest to consider alternatives. Especially hollow cathode discharge and RF glow occur promising to us. It should be feasible and would be interesting also for other applications, to find a simple circuit that can provide short high voltage RF pulses from a low DC voltage source.

3. To show the feasibility of portable operation. One important and challenging step in the development of a portable device will be the integration of a power source. Here the application of micro ionizers has the clear advantage of the possibility to be operated at comparatively low voltage. For the drift chamber it is important to choose the right drift field strength to achieve the highest resolution. This is another argument for miniaturization, since the voltage applied to the drift chamber can be lower in a shorter drift chamber.

Bon courage!

Acknowledgements

I express my sincere gratitude first of all towards my thesis supervisor Prof. Philippe Renaud who made this work possible, and who also made it pleasurable with his active support, and steady optimism and encouragement. His efforts to teach me science are gratefully acknowledged! Adding profoundly to the success of this thesis and the pleasure to work on it was Harald van Lintel, with whom I was lucky enough to share not only the office, but also some of the more significant incidents of the last years. I am happy to have found, at the Institute of Microtechnology and Microelectronics of the EPFL, the most pleasant group one can wish to work in. Thank you! I also enjoyed the cooperation with Dr. Christoph Hollenstein whose enthusiasm for the project strengthened my confidence whenever doubts threatened to strike. Thanks also to Dr. Roger Carr for being a great companion and tutor during the first steps of our venture into the vast world of ionization. I further wish to thank the jury of my thesis exam for their engagement, and especially Prof. Jean-Francis Loiseau for our enlightening and inspiring discussions. Antoine Descoeudres' execution of optical spectroscopy measurements is greatly appreciated. The students who decided to do their diploma or semester work on my thesis project, or joined me for an internship, were an invaluable support and enrichment: Ricardo Iuzzolino, Rachel Grange, Irene Vink, Ritesh Jhaveri, and Faouzi Khechana. The following organisations were vital in my research: The EPFL as a whole, and within it especially the Centre of MicroNanoTechnology (CMI), the Centre Interdepartemental de Microscopie Electronique (CIME), and the Institute of Nanostructure Physics (IPN). Amongst the supportive organizations outside the EPFL were the Paul Scherrer Institute (PSI) in Villigen, the IBM Zurich Research Laboratory, and of course "Top Nano 21" who paid for a part of my work. I also appreciate our industrial partners' support of the project: Aritron Instrumente AG (Forch), LN Industries SA (Châtelaine), Microsens SA (Neuchâtel), Orbisphères GVA (Vésenaz), GenSpec SA (Boudry). Not to forget are those who paved my way to electrical discharge science: all the authors of the books and articles from which I took profit. For proofreading the thesis I thank Harald van Lintel, Dr. Jochen Kuhnhenn, Sarah Longwitz, and (last minute) Dr. Roger Carr.

There's more to life than science: A big thank you and a hug to my family and friends. And an extra smile for my wife Sarah, who's giving meaning to it all: ☺

References

[AKI99] Akishev Y.S., Grushin M.E., Deryugin A.A., Napartovich A.P., Pan'kin M.V., Trushkin N.I., *Self-oscillations of a positive corona in nitrogen*. J. Phys. D, 1999. 32(18): pp. 2399-409

[ARK01] Arkhipenko V.I., Simonchik L.V., Sukhadolav D.V., Zgirouski S.M., *The self-sustained high current glow discharge at atmospheric pressure with normal current density*, in: Proc. Plasma 2001, Warsaw (PL), 2001

[AUB99] Aubrecht L., Koller J., Zahoranova A., *'Trichel' pulses in negative corona discharge on trees*. J. Phys. D, 1999. 32: pp. L87–L90

[BAT97] Battistoni G., et al., *Systematic study of the features of the streamer discharge by means of pulse shape analysis*. Nuclear Instruments and Methods in Physics Research A, 1997. 399: pp. 244-60

[BEC71] Beckey H.-D., *Field Ionization Mass Spectrometry*. Pergamon Press Ltd., Oxford (UK), 1971

[BIB99] Biborosch L.D., et al., *Microdischarges with plane cathodes*. Appl. Phys. Letters, 1999. 75(25): pp. 3926-8

[BOE87] Böhringer H., Fahey D.W., *Mobilities of several mass-identified positive and negative ions in air*. Int. J. Mass Spectrometry and Ion Processes, 1987. 81: pp. 45-65

[BON98] Bonard J.-M., et al., *Field emission properties of multiwalled carbon nanotubes*. Ultramicroscopy, 1998. 73: pp. 7-15

[BRA00] Braithwaite N.S.J., *Introduction to gas discharges*. Plasma Sources Sci. Technol., 2000. 9: pp. 517-27

[BRO59] Brown S.C., *Basic Data of Plasma Physics*. Wiley, New York (USA), 1959

[BUD92] Budzikiewicz H., *Massenspektrometrie - eine Einführung*. 4th ed. Wiley-VCH, Weinheim (D), 1992

[CEN99] CERN, *Townsend coefficients*: http://alice.web.cern.ch/Alice/transparencies/rjd/townsend.html (2003-11-14)

[CHO95] Cho F.Y., et al., *Ionization Gas Analyzer and Method*, Patent: US 5,475,311 (Dec. 12, 1995)

[CLE97] Clemmer D.E., Jarrold M.F., *Ion mobility measurements and their applications to clusters and biomolecules*. J. Mass Spec., 1997. 32: pp. 577-92

[COB58] Cobine J.D., *Gaseous Conductors: Theory and Engineering Applications*. Constable & Co. or Dover Publications, New York (USA), 1958

[CRA54] Craggs J.D., Meek J.M., *High Voltage Laboratory Technique*. Butterworth Scientific Publishers (now: Butterworth-Heinemann), London (UK), 1954

References

[CRC]   Lide D.R. (ed.), *CRC Handbook of Chemistry and Physics*: http://www.knovel.com/knovel/Databook/default.htm?WCI=BrowseBook&WCE=34&WCU=1 (2003-11-14)

[CRO86]   Crowley J.M., *Fundamentals of Applied Electrostatics*. John Wiley & Sons, New York (USA), 1986

[DAI99a]   Dai P., Lin G., *A physics study of a high performance micro gas sensor*, Rustosensors, New Jersey (USA), 1999

[DAI99b]   Dai P., Naruse Y., Lin G., *A Novel Micro Gas Sensor with High Sensitivity for Micro Gas Chromatograph Systems*, 1999 (Manuscript)

[DAI99c]   Dai P., Naruse Y., Lin G., *A Novel High Sensitivity Micro GC Detector*, in: Proc. Transducers '99, Sendai (J), 1999

[DAI99d]   Dai P., Lin G., *A physics study of a high performance micro gas sensor*, 1999 (Manuscript)

[DAV73]   Davies D.K., *The initiation of electrical breakdown in vacuum - a review*. J. Vac. Sci. Technol., 1973. 10(1): pp. 115-21

[DHA00]   Dhariwal R.S., Torres J.M., Desmulliez M.P.Y., *Electric field breakdown at micrometre separations in air and nitrogen at atmospheric pressure*. IEE Proceedings Science, Measurement and Technology, 2000. 147(5): pp. 261-5

[DHA94]   Dhariwal R.S., et al., *Breakdown Electric Field Strength between Small Electrode Spacings in Air*, in: Proc. Micro Systems Tech. '94, Berlin, 1994

[DHE03]   Dheandhanoo S., Ketkar S.N., *Improvement in Analysis of O2 in N2 by using Ar drift gas in an ion mobility spectrometer*. Anal. Chem., 2003. 75(3): pp. 698-700

[DYK56]   Dyke W.P., Dolan W.W., *Field emission*. Adv. Electron. Electron Phys., 1956. 3: pp. 89-185

[EIC01a]   Eiceman G.A., Tadjikov B., Krylov E., Nazarov E.G., Miller R.A., Westbrook J., Funk P., *Miniature radio-frequency mobility analyzer as a gas chromatographic detector for oxygen-containing volatile organic compounds, pheromones and other insect attractants*. J. Chromatogr. A, 2001. 917(1-2): pp. 205-17

[EIC01b]   Eiceman G.A., Nazarov E.G., Rodriguez J.E., Stone J.A., *Analysis of a drift tube at ambient pressure: Models and precise measurements in ion mobility spectrometry*. Rev. Sci. Instruments, 2001. 72(9): pp. 3610-21

[EIJ00b]   Eijkel J.C.T., Stoeri H., Manz A., *A dc microplasma on a chip employed as an optical emission detector for gas chromatography*. Anal. Chem., 2000. 72(11): pp. 2547-52

[EIJ00c]   Eijkel J.C.T., Stoeri H., Manz A., *An atmospheric pressure dc glow discharge on a microchip and its application as a molecular emission detector*. J. Anal. At. Spectrom., 2000. 15(3): pp. 297-300

[EIJ99]   Eijkel J.C.T., et al., *A molecular emission detector on a chip employing a direct current microplasma*. Anal.Chem., 1999. 71(14): pp. 2600-6

References

[ERC01] Ercilbengoa A.E., Spyrou N., Loiseau J.F., *Anodic glow and current oscillations in medium- and low- pressure dark discharges.* 2001. 34(4): pp. 584-592

[ERC99] Ercilbengoa A.E., Thesis: *Etude expérimentale des régimes de décharge continue positive dans l'azote et l'air pour différentes pressions*, Pau: Université de Pau et des Pays de l'Adour, 1999

[FOE96] Foerster J.A., Thesis: *Integrated Micro Vacuum Tubes in Silicon*, Delft University of Technology, Delft: Delft University Press, 1996

[FOW28] Fowler R.H., Nordheim L., *Electron Emission in Intense Electric Fields.* P. Roy. Soc. A, 1928. CXIX.-A.: pp. 173-81

[FRA97] Frame J.W., et al., *Microdischarge devices fabricated in silicon.* Appl. Phys. Lett., 1997. 71(9): pp. 1165-7

[FRE95] Freidhoff C.B., *Solid State Micro-Machined Mass Spectrograph Universal Gas Sensor*, Patent: US 5,386,115 (Jan. 31, 1995)

[FRE96] Freidhoff C.B., Young R.M., *Miniaturized mass filter*, Patent: US 5,536,939 (Jul. 16, 1996)

[FRE99a] Freidhoff C.B., Young R.M., et al., *Chemical Sensing Using Non-Optical MEMS.* JVST A, 1999: pp. 1-31

[FRE99b] Freidhoff C.B., et al., *Chemical sensing using nonoptical microelectromechanical systems.* J. Vac. Sci. Technol. A, 1999. 17 (4), Part 2: pp. 2300-7

[GEM32] Gemant A., Philippoff v., *Die Funkenstrecke mit Vorkondensator.* Ztschr. f. tech. Physik, 1932. 13: pp. 425-30

[GEO01] Georghiou G.E., Morrow R., Metaxas A.C., *The effect of photoemission on the streamer development and propagation in short uniform gaps.* J. Phys. D, 2001. 34(2): pp. 200-8

[GER59] Germer L.H., *Electrical breakdown between close electrodes in air.* J. Appl. Phys., 1959. 30(1): pp. 46-51

[GES00] Gessner C., Scheffler P., Gericke K.-H., *Characterisation of a novel micro-structured plasma source via optical emission and laser induced fluorescence spectroscopy*, Braunschweig (D), 2000 (Manuscript)

[GOL03] Golota V., Zavada L., Kadolin B., Karas' V., Paschenko I., Pugach S., Yakovlev A., *Investigation of nonstationary modes of atmospheric pressure needle-to-plane gas discharge and streamer propagation*, in: Proc. Int. Conf. on Phenomena in Ionized Gases, Greifswald (D), 2003

[GOM61] Gomer R., *Field emission and field ionization.* Harvard University Press, Cambridge (UK), 1961

[HAG00] Hagelaar G.J.M., Thesis: *Modeling of microdischarges for display technology*, Eindhoven (NL): TU Eindhoven, 2000

[HEL97] Held B., et al., *Self-Sustained Conditions in Inhomogeneous Fields.* J. Phys. III France, 1997. 7: pp. 2059-77

References

[HOP00] Hopwood J.A., *A microfabricated inductively coupled plasma*. J. MEMS, 2000. 9(3): pp. 309-13

[http01] Stoecki, http://dsa.mpi-muelheim.mpg.de/kofo/stoecki/ -obsolete-
try: http://www.mpi-muelheim.mpg.de/kofo/institut/forschungsinfrastruktur/ms_schrader/ms_group.html

[http06a] Tissue B.M., http://www.chem.vt.edu/chem-ed/ms/ms-intro.html (2003-11-10)

[JEN01] Jenion, *Radio-frequency plasmas*:
http://www.jenion.de/html/atmospheric_pressure_plasma.html -obsolete- (2001-09-18)

[KHA01] Khayamian T., Tabrizchi M., Taj N., *Direct determination of ultra-trace amounts of acetone by corona-discharge ion mobility spectrometry*. Fresenius J. Anal. Chem., 2001. 370(8): pp. 1114-6

[KHA03] Khayamian T., Tabrizchi M., Jafari M.T., *Analysis of 2,4,6-trinitrotoluene, pentaerythritol tetranitrate and cyclo-1,3,5-trimethylene-2,4,6-trinitramine using negative corona discharge ion mobility spectrometry*. Talanta, 2003. 59(2): pp. 327-33

[KIM01] Kim Y.-K., Irikura K.K., Rudd M.E., Zucker D.S., Zucker M.A., Coursey J.S., Olsen K.J., Wiersma G.G., *Electron-Impact Ionization Cross Sections*:
http://physics.nist.gov/PhysRefData/Ionization/EII_table.html (2003-10-22)

[KIS59] Kisliuk P., *Electron emission at high fields due to positive ions*. J. Appl. Phys., 1959. 30: pp. 51-5

[KOR00] Kornienko O., Reilly P.T.A., Whitten W.B., Ramsey J.M., *Field-emission cold cathode EI source for a microscale ion trap mass spectrometer*. Anal. Chem., 2000. 72(3): pp. 559-62

[KOR99] Kornienko O., et al., *Electron impact ionization in a micro ion trap mass spectrometer*. Rev. Sci. Instruments, 1999. 70(10): pp. 3907-9

[KOT96] Kotvas J., Braggins T., Young R.M., Freidhoff C.B., *Method for manufacturing a miniaturized solid state mass spectrograph*, Patent: US 5,492,867 (Feb. 20, 1996)

[LAT81] Latham R.V., et al., *High Voltage Vacuum Insulation*, ed. R.V. Latham. Academic Press, London (UK), 1981

[LIN97] Lin G., *Solid-state Gas Sensors*, Patent: US 5,876,314 (Jan. 7, 1997)

[LIU95] Liu C.J., Rhee M.-J., *Experimental investigation of breakdown voltage characteristics of single-gap and multigap pseudosparks*. IEEE Trans. Plasma Sci., 1995. 23(3): pp. 235-8

[LOL52] Loeb L.B., *Secondary processes active in the electrical breakdown of gases*. British journal of applied physics, 1952. 3: pp. 341-9

[LON01] Longwitz R.G., van Lintel H., Carr R., Hollenstein C., Renaud P., *Study of gas ionization schemes for micro devices*, in: Proc. Transducers '01, München (D), 2001

[LUX01] Lux J., *Paschen's Law*: http://home.earthlink.net/~jimlux/hv/paschen.htm (2003-10-20)

[MAD93] Madou M.J., Morrison S.R., *High-Field Operation of Submicrometer Devices at Atmospheric Pressure.* Abstracts of papers of the American Chemical Society, 1993. 206: 185-COLL, Part 1: pp. 145-9

[MAR98] Martin S.J., et al., *Ion mobility spectrometer using frequency-domain separation,* Patent: US 5,789,745 (Aug. 04, 1998)

[MIL67] Miller H.C., *Change in field intensification factor beta of an electrode projection (whisker) at short gap lengths.* J. Appl. Phys., 1967. 38(11): pp. 4501-4

[MIR01] Miller R.A., Nazarov E.G., Eiceman G.A., Thomas King A., *A MEMS radio-frequency ion mobility spectrometer for chemical vapor detection.* Sensors and Actuators A, 2001. 91(3): pp. 307-18

[MIR02] Miller R.A., Zapata A., Nazarov E.G., Krylov E., Eiceman G.A., *High performance micromachined planar field-asymmetric ion mobility spectrometers for chemical and biological compound detection,* in *Biomems and Bionanotechnology.* 2002. p. 139-47

[MIR99] Miller R.A., Eiceman G.A., Nazarov E.G., *A Micromachined Field Asymmetric-Ion Mobility Spectrometer (Fa-Ims),* in: Proc. 8th International Conference on Ion Mobility Spectrometry, Buxton, Derbyshire (UK), 1999

[MOE00] Möller D., Güntherodt H.-J., Haefke H., *Nanoplasma - Production, characterization and application of a mesoscopic plasma.* NR NPR 36, 2000: pp. 82-83

[MOE99a] Möller D., et al., *High-pressure STM study of microscopic gas discharges.* Surf. Interface anal., 1999. 27: pp. 525-9

[MOE99b] Möller D., Thesis: *Erzeugung, Charakterisierung und Anwendung Mesoskopischer Gasentladungen mit einem Hochdruck-Rastertunnelmikroskop,* Basel (CH): Universität Basel, 1999

[MOR97a] Morrow R., *The theory of positive glow corona.* J. Phys. D, 1997. 30(22): pp. 3099-114

[MOR97b] Morrow R., Lowke J.J., *Streamer propagation in air.* J. Phys. D, 1997. 30: pp. 614-27

[MUE01] Müller J., *HF-Plasma-Gasdetektor als Mikrosystem*: http://www.tu-harburg.de/mst/deutsch/forschung/muel_06.shtml (2003-08-08)

[MUL37] Müller E.W., Z. Physik, 1937. 106: pp. 541-50

[NAI99] Naidu M.S., Kamaraju V., *High Voltage Engineering.* 2 ed. McGraw Hill, 1999

[NAS71] Nasser E., *Fundamentals of Gaseous Ionization and Plasma Electronics.* Wiley, New York (USA), 1971

[ONO00] Ono T., Youn Sim D., Esashi M., *Micro-discharge and electric breakdown in a microgap.* J. Micromech. Microeng., 2000. 10(3): pp. 445-51

[ONO00b] Ono T., Dong Youn S., Esashi M., *Imaging of micro-discharge in a micro-gap of electrostatic actuator,* in: Proc. IEEE 13th Ann. Int. Conf. MEMS, Miyazaki (J), 2000

[OSM94] Osmokrovic P., et al., *Mechanism of Electrical Breakdown Left of Paschen Minimum.* IEEE Trans. Dielectr. Electr. Insul., 1994. 1(1)

References

[PAS1889] Paschen F., *Über die zum Funkenübergang in Luft, Wasserstoff and Kohlensäure bei verschiedenen Drücken erforderliche Potentialdifferenz.* Weid. Annalen der Physick, 1889. 37: pp. 69-75

[PAT02] Patsch R., Berton F., *Pulse sequence Analysis - a diagnostic tool based on the physics behind partial discharges.* J. Phys. D, 2002. 35: pp. 25-32

[PEP97] Pedrow P., Olsen R., *Peek's law*:
http://www.eecs.wsu.edu/~pedrow/HV_Engineering/lecture/bdgas/part2/ (2003-07-14)

[PET00] Petzold G., Siebert P., Müller J., *A Micromachined Electron Beam Ion Source*, in: Proc. MicroTAS, Twente (NL), 2000

[RAI87] Raizer Y.P., *Gas discharge physics.* 1987 ed. Springer, Berlin, 1987 (recent ed.: 1997)

[REE97] Reess T., Paillol J., *The role of the field-effect emission in Trichel pulse development in air at atmospheric pressure.* J. Phys. D, 1997. 30: pp. 3115-22

[ROB68] Roboz J., *Introduction to Mass Spectrometry.* Interscience Publishers, New York (US), 1968

[ROT00] Roth D., Schlemm H., Gericke K.H., Schmidt-Böcking H., *Anordnung zur grossflaechigen Erzeugung von Hochfrequenz-Niedertemperatur-Plasmen bei Atmosphärendruck*, Patent: DE 10032955 A1 (2000)

[SCK95] Schoenbach K.H., et al., *Microhollow cathode discharges.* Appl. Phys. Lett., 1995. 68(1): pp. 13-5

[SCK97] Schoenbach K.H., et al., *High-pressure hollow cathode discharges.* Plasma Sources Sci. Technol., 1997. 6: pp. 468-77

[SCM01] Schlemm H., Roth D., *Atmospheric pressure plasma processing with microstructure electrodes and microplanar reactors.* Surf. and Coatings Technology, 2001. 142-144: pp. 272-6

[SCP00] Scheffler P., Gessner C., Gericke K.-H., *Micro-structured electrode arrays: a new discharge device for pollution control*, Braunschweig (D), 2000 (Manuscript)

[SED96] Sedlacek M., *Electron physics of vacuum and gaseous devices.* John Wiley & Sons, Inc., New York (USA), 1996

[SEL97] Selwyn G., *Atmospheric pressure plasma cleaning of contamination surfaces*: http://www-emtd.lanl.gov/TD/science/AtmosphericPlasmaCleaning.html (2003-11-14)

[SHU86] Schumate C., St. Louis R.H., H.H. Hill J., *Table of reduced mobiliity values from ambient pressure ion mobility spectrometry.* J. Chromatography, 1986. 373: pp. 141-73

[SIE98] Siebert P., Petzold G., Müller J., *Surface microstructure/miniature mass spectrometer: processing and applications.* Appl. Phys. A, 1998. 67(2): pp. 155-60

[SIE99] Siebert P., Petzold G., Müller J., *Processing of complex microsystems: A Micro Mass Spectrometer.* SPIE, 1999. 3680: pp. 562-571

[SIL01] Sillon N., Baptist R., *Micromachined mass spectrometer*, in: Proc. Proc. Transducers '01, München (D), 2001

[SIL02]   Sillon N., Baptist R., *Micromachined mass spectrometer.* Sensors and Actuators B, 2002. 83: pp. 129-137
[SIT95]   Sittler F., *Micromachined mass spectrometer*, Patent: US 5,401,963 (Mar. 28, 1995)
[SIT96]   Sittler F., *Micromachined mass spectrometer*, Patent: US 5,541,408 (Jul. 30, 1996)
[SIW94]   Siems W.F., Wu C., Tarver E.E., Hill H.H., Larsen P.R., McMinn D.G., *Measuring the resolving power of ion mobility spectrometers.* Anal. Chem., 1994. 66: pp. 4195-201
[SPA86]   Spangler G.E., Vora K.N., Carrico J.P., *Miniature ion mobility spectrometer cell.* J. Phys. E, 1986. 19(3): pp. 191-8
[SPA93]   Spangler G.E., *Theory and technique for measuring mobility using ion mobility spectrometry.* Anal. Chem., 1993. 65(21): pp. 3010-4
[SPA99]   Spangler G.E., *The effect of cluster thermochemistry on the energetics of an ion in ion mobility spectrometry*, in: Proc. 8th Int. Conf. Ion Mobility Spectrometry, Buxton (UK), 1999
[SPY95]   Spyrou N., Peyrous R., Soulem N., Held B., *Why Paschen's law does not apply in low-pressure gas discharges with inhomogeneous fields.* J. Phys. D, 1995. 28(4): pp. 701-10
[STA96]   Stalder R.E., et al., *Array of micro-machined mass energy micro-filters for charged particles*, Patent: US 5,486,697 (Jan. 23, 1996)
[STR99a] Stark R.H., Schoenbach K.H., *Direct current glow discharges in atmospheric air.* Appl. Phys. Letters, 1999. 74(25): pp. 3770-2
[SYM98]   Syms R.R.A., Tate T.J., et al., *Design of a microengineered electrostatic quadrupole lens.* IEEE Trans. Electr. Dev., 1998. 45(11): pp. 2304-11
[TAB00]   Tabrizchi M., Khayamian T., Taj N., *Design and optimization of a corona discharge ionization source for ion mobility spectrometry.* Rev. Sci. Instrum., 2000. 71(6): pp. 2321-8
[TAB99]   Tabrizchi M., Khayamian T., Taj N., *Corona Discharge Ion Mobility Spectrometry*, in: Proc. 8th Int. Conf. Ion Mobility Spectrometry, Buxton (UK), 1999
[TAY01]   Taylor S., Tindall R.F., Syms R.R.A., *Silicon based quadrupole mass spectrometry using microelectromechanical systems.* J. Vac. Sci. Technol., 2001. 19(2): pp. 557-62
[TAY03]   Taylor S., Gibson J.R., Srigengan B., *Miniature mass spectrometry: implications for monitoring of gas discharges.* Sensor Review, 2003. 23(2): pp. 150-4
[TAY98a] Taylor S., Tunstall J.J., Syms R.R.A., Tate T.J., Ahmad M.M., *Silicon micromachined mass filter for low power, low cost quadrupole mass spectrometer*, in: Proc. IEEE Int. Workshop MEMS '98, Heidelberg (D), 1998
[TAY98b] Taylor S., Tunstall J.J., Syms R.R.A., Tate T.J., Ahmad M.M., *Initial results for a quadrupole mass spectrometer with a silicon micromachined mass filter.* IEEE Electr. Letters, 1998. 34(6): pp. 546-7
[TEE01]   Teepe M., Baumbach J.I., Neyer A., Schmidt H., Pilzecker P., *Miniaturized ion mobility spectrometer.* IJIMS, 2001. 4(1): pp. 60-4

References

[TEL02]   Telrandhe M., Peddanenikalva H., Bhansali S., Short R.T., *Microfabricated cylindrical ion trap mass spectrometer microarrays for portable analysis*, in: Proc. Eurosensors '02, Prague (CZ), 2002

[TER96]   Terashima K., Howald L., Haefke H., Güntherodt H.-J., *Develpoment of a Mesoscale/Nanoscale Plasma Generator*. Thin Solid Films, 1996. 281-282: pp. 634-6

[TOR99a]   Torres J.-M., Dhariwal R.S., *Electric field breakdown at micrometre separations*. Nanotechnology, 1999. 10(1): pp. 102-7

[TOR99b]   Torres J.-M., Dhariwal R.S., *Electric field breakdown at micrometre separations in air and vacuum*. Microsyst. Tech., 1999. 6: pp. 6-10

[TSU01]   Tsui Y.Y., *Survey of plasma concepts and applications*, 2001 (http://www.ee.ualberta.ca/~ee583/ch1.pdf) (Manuscript)

[TUN98]   Tunstall J.J., Taylor S., Syms R.R.A., Tate T., Ahmad M.M., *Silicon micromachined mass filter for a low power, low cost quadrupole mass spectrometer*, in: Proc. MEMS '98, Heidelberg (D), 1998

[VEL01]   van Veldhuizen E.M., Rutgers W.R., *Corona discharges: fundamentals and diagnostics*, in: Proc. Workshop on frontiers in low temperature plasma diagnostics IV, Limburg (NL), 2001

[VEL02]   van Veldhuizen E.M., Rutgers W.R., *Pulsed positive corona streamer propagation and branching*. J. Phys. D., 2002. 35: pp. 2169-79

[WAS72a]   Wasserrab T., *Gaselektronik I*. Bibliographisches Institut, Mannheim, 1972

[WAS72b]   Wasserrab T., *Gaselektronik II*. Bibliographisches Institut, Mannheim, 1972

[WHI01]   Whitten W.B., Ramsey J.M., et al., *Microchip ion trap mass spectrometry*, in: Proc. MicroTAS, Enschede (NL), 2001

[WIB00]   Wiberg D.V., et al., *A LIGA fabricated quadrupole array for mass spectroscopy*, 2000 (Manuscript)

[WIB01]   Wiberg D.V., *Toward a micro gas chromatograph/mass spectrometer system*, 2001 (Manuscript)

[WIK03]   Wikipedia, *Wikipedia*: http://www.wikipedia.org/ (2003-11-14)

[XU00]   Xu J., Whitten W.B., Ramsey J.M., *Space charge effects on resolution in a miniature ion mobility spectrometer*. Anal. Chem., 2000. 72(23): pp. 5787-91

[XU01]   Xu J., Whitten W.B., Ramsey J.M., *Miniature ion mobility spectrometer detector with a pulsed ionization source*, in: Proc. MicroTAS, Enschede (NL), 2001

[XU98]   Xu J., Kung C.Y., Whitten W.B., Ramsey J.M., *Studies of Miniature Ion Mobility Spectrometer*, in: Proc. 7th Int. Workshop Ion Mobility Spectrometry, Hilton Head Island (USA), 1998

[XU99]   Xu J., Whitten W.B., Ramsey J.M., *Miniature Ion Mobility Spectrometry*, in: Proc. 8th Int. Conf. Ion Mobility Spectrometry, Buxton (UK), 1999

[YAS01]   Yasuoka K., et al., *Pulsed operation of micro-hollow cathode plasma*, Tokyo Inst. of Tech., 2001 (Manuscript)

[YOO01a] Yoon H.J., et al., *The Fabrication of the Novel Micro Time-of-Flight (TOF) Mass Spectrometer with a Micro Ion Source*, in: Proc. Transducers '01, München (D), 2001

[YOO01b] Yoon H.J., Kim J.H., Park T.G., Yang S.S., Jung K.W., *The test of hot electron emission for the micro mass spectrometer*, in *Design,Test Integration, and Packaging of MEMS/MOEMS*. 2001. pp. 360-7

[YOU96] Young R.M., Freidhoff C.B., *Miniature mass spectrograph separation embodiments*, Patent: US 5,536,939 (Nov. 16, 1996)

[ZIM01] Zimmermann S., Krippner P., Vogel A., Müller J., *Miniaturized Flame Ionization Detector for Gas Chromatography*, in: Proc. Transducers '01, München (D), 2001

Appendix

A. Signs, units, constants, and abbreviations

In general, SI units are used throughout, see IEEE standard 945-1984, *Recommended Practice for Preferred Metric Units for Use in Electrical and Electronics Science and Technology*. Table 17 is a list of the units that were used in this thesis.

Table 17 List of quantities and their units that were used in the thesis.

Sign	Explanation	Unit/Value
α	first ionization (Townsend) coefficient	1/m
β	field intensification/amplification factor	1
μ	Fermi level; mobility	eV; m²/Vs
μ_e	mobility of electrons	m²/Vs
$\mu_{+/-}$	mobility of positive/negative ions	m²/Vs
ν	frequency; particle flux	1/s; m⁻²s⁻¹
ν_m	collision frequency (ion mobility), effective collision frequency for momentum transfer	1/s
σ	conductivity; cross section	S/m=1/Ωm; m²
σ_e	cross section of electron/particle collision	m²
σ_m	cross section of particle/particle collision	m²
τ	time constant	s
Γ	flux	1/m²s
γ	second ionization coefficient	1
Γ	flux	1/m²s
Φ	work function; electric potential	eV; V
ε_0	permittivity of vacuum	8.8542×10⁻¹² As/Vm
ε_r	relative permittivity (formerly called the dielectric constant)	1
η	overvoltage	V
κ, ε_r	dielectric constant, normalised permittivity (=D/E)	1
λ	length of mean free path	m
λ_D	Debye length	m
L	characteristic length	m
ρ	charge density, space charge; mass density	C/m³; g/m³

Appendix

Sign	Explanation	Unit/Value
d	gap distance, diameter	m
A	area; mass number	m^2; g/mol
a	acceleration	m/s^2
C	capacitance	F (farad)
C_{gc}	grid-cathode capacitance	F
d	distance	m
d_m	mean distance between two molecules	m
D	electric displacement; tunnelling probability; diffusion coefficient	As/m^2; 1; m^2/s
e	electron charge	1.6022×10^{-19} As
$\bar{e}$	natural logarithm number	2.7183
E	electric field strength	V/m
E_a	field from applied voltage	V/m
E_d	drift field strength	V/m
E^*	atomic excitation energy	J
E_x	kinetic energy along the emission direction; extinction field strength	J; V/m
E_i	initiation field strength	V/m
F	force	N
G	conductance	$S = 1/\Omega$
$h, \hbar$	Planck's constant	6.63×10^{-34} Js
I, IP, V_i	ionization (appearance) potential/energy	eV
I	current	A = C/s
J	glow intensity	1
j	current density	A/m^2
k	Boltzmann constant; form/degradation factor	1.4×10^{-23} JK^{-1}; 1
l	length	m
l_d	drift length	m
L	characteristic dimension, distance between electrodes; inductance	m; H
m	mass	g
m_e	electron rest mass	9.1094×10^{-28} g
m/e	mass to charge ratio	g/C
n	number density	$1/m^3$
n_0, n_a	density of neutral gas	$1/m^3$
n_e	density of electrons	$1/m^3$
n_j	density of ions	$1/m^3$
$\dot{n}$	number of particles arriving in a volume element	1

Appendix

Sign	Explanation	Unit/Value
N	number of gas molecules; plate number	1
N_A	Avogadro's number	6.02×10^{23} mol^{-1}
N_L	Loschmidt number	2.6873×10^{19} cm^{-3}
P	gas pressure; power density	Pa; W/m^3
P_c	number of collisions per m or inverse free path length [RAI87] $p10$	m^{-1}Pa^{-1}
Q	charge	C
Q_s	surface charge	C/m^2
R	resistance; resolution; universal gas constant	Ω; 1; 8.3145 J/molK
r	radius, radius of deflection	m
r_k	gas kinetic radius	m
T	(thermodynamic) temperature	°C, K
T_e	electron temperature	K
t	time	s
t_d	drift time	s
t_{iz}	ionization pulse time	s
t_g	gate time	s
TI	total ion current	A
v_d	drift velocity	m/s
u, amu	atomic mass unit	1.67×10^{-24} g
V	voltage, electric potential, electromotive force; volume	V; m^3
V_c	(corona) ignition voltage	V
V_i	ionization potential	eV
V_d	drift voltage	V
V_x	lower voltage limit for discharge: extinction voltage	V
V_b	breakdown voltage	V
V_t	transition voltage (to self sustained discharge)	V
v	(directed) velocity	m/s
v_{iT}	thermal velocity of an ion	m/s
v_m	mean particle voltage	m/s
w	(undirected) velocity; electrode width; peak width	m/s; m; s
$w_{1/2}$	full width at half-maximum of a detected ion peak	s
W	work, energy	J
x	weighting factor; general purpose variable	1; any

Appendix

Unit conversions

1 bar = 10^5 Pa (1 mbar = 1 hPa); 1 Torr = 1 mmHg = 133.32 Pa; 1 Pa = 7.5×10^{-3} Torr; 1 cmTorr = 1.3332 Pa×m

Abbreviations

Table 18 List of abbreviations that are used in the thesis.

Abbr.	Meaning
EI	electron impact ionization
EMF	electromotive force
ESI	electrospray ionization
FAIMS	High-Field Asymmetric waveform Ion Mobility Spectrometer
FEE	field electron emission
FI	field ionization
F-N	Fowler-Nordheim
IMS	ion mobility spectrometry / spectrometer
LIGA	LItographie (lithography), Galvanoformung (electroplating) and Abformung (injection molding)
MEMS	micro electromechanical system
MS	mass spectrometry
OPAMP	operational amplifier
PCB	printed circuit board
RF	radio frequency
UV	ultraviolet (light)

B. Properties of gases

Table 19 Ionization potential and mean free path of various gases at 300 K and 10^5 Pa.

Gas	ionization potential V_i [eV]	mean free path λ at V_i [µm]	electron affinity** [eV]
He	24.6 (highest of any gas) */**	1.56*	–
CO_2	13.8**		C: 2.55
Ar	15.8**		–

Gas	ionization potential V_i [eV]	mean free path λ at V_i [μm]	electron affinity** [eV]
Ne	21.6**		–
CH_4	12.6**		
O_2	12.5*, 12.1**	0.40*	0.4510 ± 0.0070 Travers, Cowles, et al., 1989 (NIST) O: 3.44
N_2	15.6*/**	0.35*	N: 3.04
H_2	15.4*/**	0.52*	H: 2.2
H_2O	12.7*, 12.6**	0.20*	H_2O^-: 1.20 ± 0.50 Griffiths and Harris, 1987 (NIST)

*[FOE96], **[CRC]

C. List of used instruments

Pressure meters

Pirani gauge: Edwards APG-M, Active gauge (Penning): Edwards AIM-S
Membrane gauge MP 330

Power supplies

Heinzinger PNC 3500-50 ump: Range of 0-3500 V
Witmer TF 300/0.5
Batteries

Current and voltage measurement

Multimeter Metex M-4460, Voltcraft VC 444, Hewlett Packard 34401A
Picoampere amplifier: Stanford Research Systems SR570
Flow rates were measured with Aalborg flow meters

Oscilloscope Agilent 54621A

Internal resistance: 1 MΩ, bandwidth: 60 MHz, sampling rate: 200 MHz

Appendix

Other equipment

Resistors: Carbon and metal strip types
Operational amplifier: OPA128
Chip holder: SOP 16p socket from Wells Japan Ltd.

D. Vacuum system

The vacuum system was set up on an Edwards Pico Dry vacuum pump system with integrated controller for the turbo molecular pump and several pressure gauges.

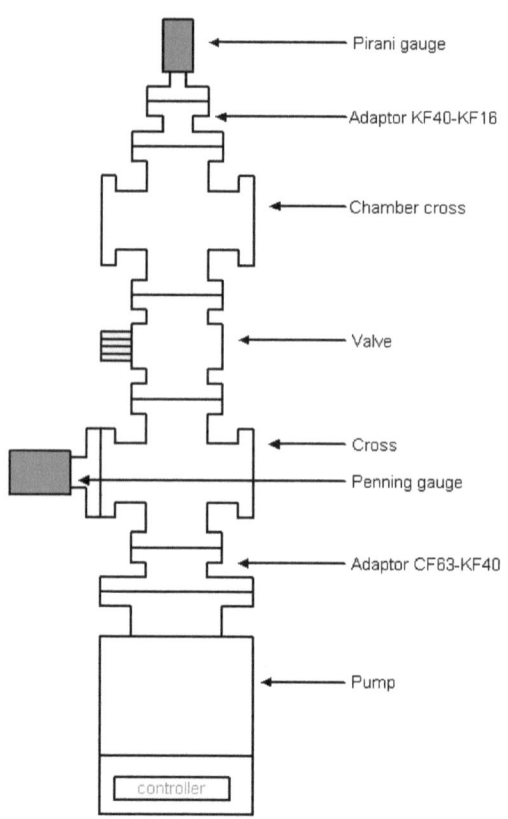

Fig. 79 Schematic drawing of the original design of the vacuum system used for most of the micro ionizer experiments in a controlled atmosphere.

Appendix

E. Pulse generation circuits

Pulse generator I

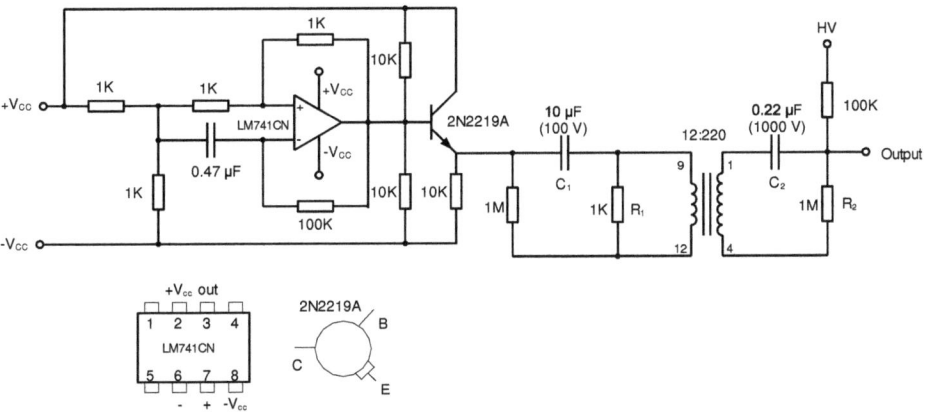

Fig. 80 Pulse-generator circuit, 1st version. 1: OPAMP multivibrator, 2: Emitter follower (output buffer), 3: Differentiator 1, 4: Pulse voltage amplification through a transformer, 5: Differentiator 2.

Pulse generator II

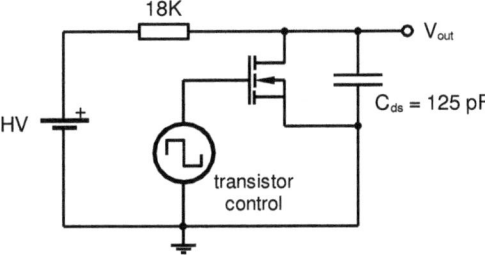

Fig. 81 HV pulse generator, 2nd version. The parasite capacity C_{ds} limits the rise time of the pulses. Transistor: MOS 2SK1317.

Appendix

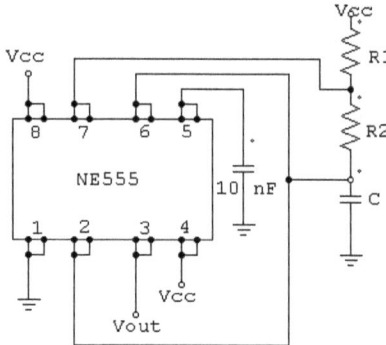

Fig. 82 Command circuit for the transistor in the circuit of Fig. 81. The command circuit is based on a multivibrator of type NE555. The resistors R1 and R2, and the capacitor C determine the signal characteristics.

F. Micro ionizer processes

Type Plan

Table 20 The principal Plan process steps.

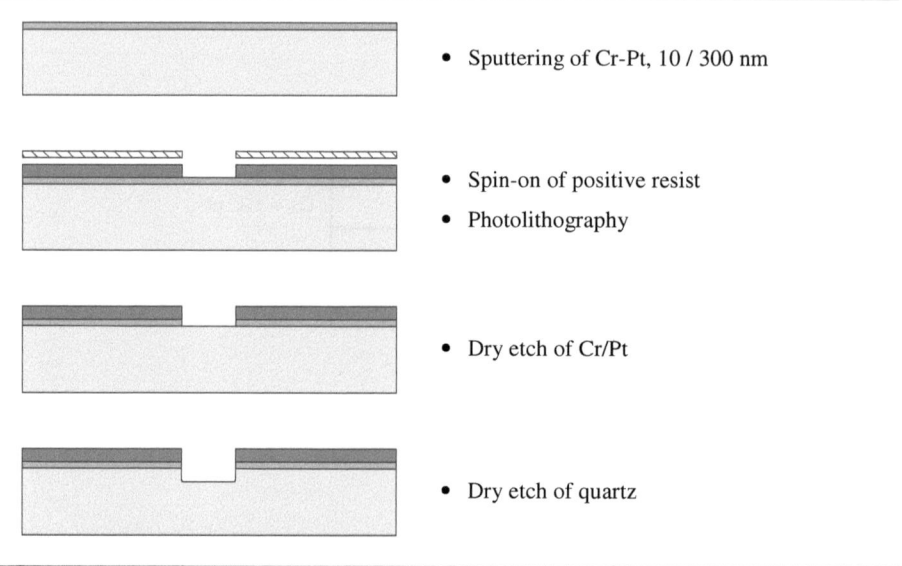

- Sputtering of Cr-Pt, 10 / 300 nm

- Spin-on of positive resist
- Photolithography

- Dry etch of Cr/Pt

- Dry etch of quartz

VIII

Appendix

Type I-Plan

Table 21 The principal I-Plan process steps.

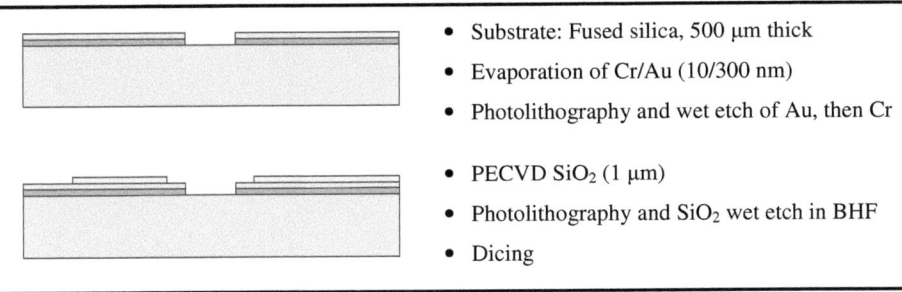

	• Substrate: Fused silica, 500 µm thick
	• Evaporation of Cr/Au (10/300 nm)
	• Photolithography and wet etch of Au, then Cr
	• PECVD SiO$_2$ (1 µm)
	• Photolithography and SiO$_2$ wet etch in BHF
	• Dicing

Type Si-Bulk

Table 22 The principal Si-Bulk process steps.

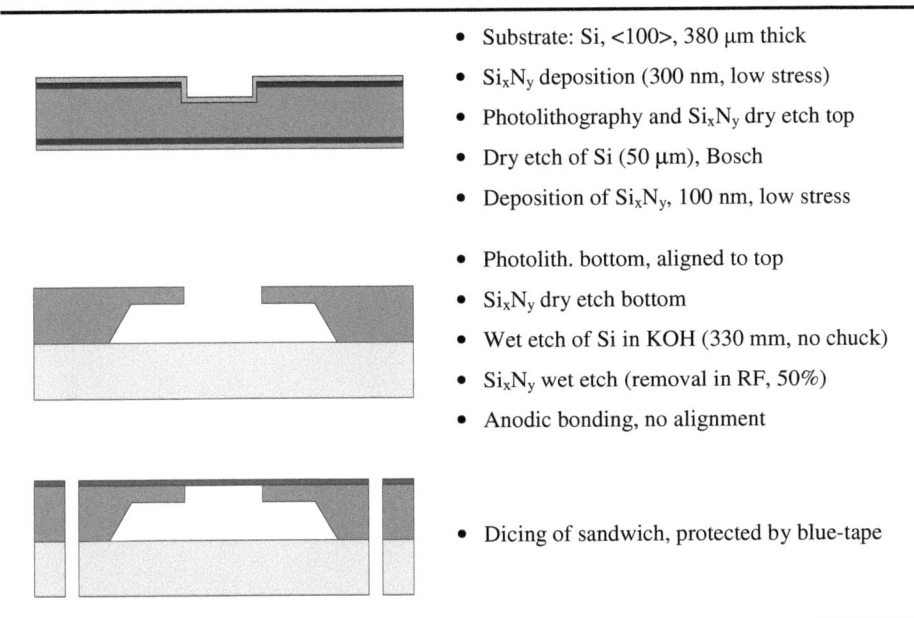

	• Substrate: Si, <100>, 380 µm thick
	• Si$_x$N$_y$ deposition (300 nm, low stress)
	• Photolithography and Si$_x$N$_y$ dry etch top
	• Dry etch of Si (50 µm), Bosch
	• Deposition of Si$_x$N$_y$, 100 nm, low stress
	• Photolith. bottom, aligned to top
	• Si$_x$N$_y$ dry etch bottom
	• Wet etch of Si in KOH (330 mm, no chuck)
	• Si$_x$N$_y$ wet etch (removal in RF, 50%)
	• Anodic bonding, no alignment
	• Dicing of sandwich, protected by blue-tape

Die VDM Verlagsservicegesellschaft sucht für wissenschaftliche Verlage abgeschlossene und herausragende

Dissertationen, Habilitationen, Diplomarbeiten, Master Theses, Magisterarbeiten usw.

für die kostenlose Publikation als Fachbuch.

Sie verfügen über eine Arbeit, die hohen inhaltlichen und formalen Ansprüchen genügt, und haben Interesse an einer honorarvergüteten Publikation?

Dann senden Sie bitte erste Informationen über sich und Ihre Arbeit per Email an *info@vdm-vsg.de*.

Sie erhalten kurzfristig unser Feedback!

VDM Verlagsservicegesellschaft mbH
Dudweiler Landstr. 99   Telefon +49 681 3720 174
D - 66123 Saarbrücken   Fax   +49 681 3720 1749
www.vdm-vsg.de

Die VDM Verlagsservicegesellschaft mbH vertritt

Printed by Books on Demand GmbH, Norderstedt / Germany